SEX, IDENTITY AND HERMAPHRODITES
IN IBERIA, 1500–1800

The Body, Gender and Culture

Series Editor: *Lynn Botelho*

FORTHCOMING TITLES

SEX, IDENTITY AND HERMAPHRODITES IN IBERIA, 1500–1800

BY

Richard Cleminson and Francisco Vázquez García

Routledge
Taylor & Francis Group

LONDON AND NEW YORK

First published 2013 by Pickering & Chatto (Publishers) Limited

Published 2016 by Routledge
2 Park Square, Milton Park, Abingdon, Oxfordshire OX14 4RN
711 Third Avenue, New York, NY 10017

First issued in paperback 2015

Routledge is an imprint of the Taylor & Francis Group, an informa business

BRITISH LIBRARY CATALOGUING IN PUBLICATION DATA

Cleminson, Richard, author.
Sex, identity and hermaphrodites in Iberia, 1500–1800. – (The body, gender and culture)
1. Intersexuality – Iberian Peninsula – History. 2. Gender identity – Iberian Peninsula – History.
I. Title II. Series III. Vázquez García, Francisco, author.
306.7'685'0946-dc23

ISBN-13: 978-1-138-66459-3 (pbk)
ISBN-13: 978-1-8489-3302-6 (hbk)

Typeset by Pickering & Chatto (Publishers) Limited

CONTENTS

ACKNOWLEDGEMENTS

We would like to thank Juan Martos Fernández of the Department of Greek and Latin Philology at the University of Seville for his generous assistance in translating many of the medical texts consulted in order to produce this work. Our thanks also go to María Jesús Ruiz and José María Fraile for directing us towards relevant sources, and to Sonia Muñoz-Prián for allowing us to cite her Master's dissertation.

Francisco Vázquez García would like to record his thanks to his colleagues in the Department of Philosophy at the University of Cadiz, namely Ramón Vargas-Machuca, José Luis Moreno Pestaña, Juan Carlos Mougán Rivero, Jesús González Fisac, Antonio Frías and Cándido Martín, for providing such a pleasant and supportive atmosphere in which to work. He would also like to thank Oliva and Curro, whose constant presence and love have meant that his spirits have remained constant over the period it has taken to research and write this book.

Richard Cleminson wishes to thank his colleagues at the University of Leeds, both in the School of Modern Languages and Cultures and in the Centre for Interdisciplinary Gender Studies, for their interest and support and for providing an excellent working environment. Thanks also go to the staff at Pickering & Chatto who accompanied this project at its inception, namely Ruth Ireland, and Mark Pollard and Janka Romero in its later stages. Encouragement to continue looking for new ideas and interpretations has been provided over the years by Professors Chris Perriam, University of Manchester, and Alison Sinclair, University of Cambridge; and Dr Rosa María Medina Doménech, University of Granada, has, in many senses, been a collaborator in this and other projects focusing on the history of sexuality in Spain. My garden has been a source of tranquil reflection and inspiration; Fredy Vélez Rodríguez has provided love and care.

Some sections of this book draw on previously published material, including our *Hermaphroditism, Medical Science and Sexual Identity in Spain, 1850–1960* (Cardiff: University of Wales Press, 2009), especially ch. 1, and 'Subjectivities in Transition: Gender and Sexual Identities in Cases of "Sex Change" and "Hermaphroditism" in Spain, c. 1500–1800', *History of Science*, 48:159 (2010), pp. 1–38.

INTRODUCTION: SEX, GENDER AND HISTORICITY

In a short article published in 1928 in the highly specialist medico-legal journal *Archivo de Medicina Legal*, the forensic doctor Asdrúbal António de Aguiar provided what was one of the first overviews on hermaphroditism in early twentieth-century Portuguese medical circles. De Aguiar's discussion and reproduction of documentation relevant to a long-lost case of suspected hermaphroditism from 1622 at the Convent of Santa Cruz in the small Alentejo town of Vila Viçosa was preceded by a historical and diagnostic framework that would structure the author's future work on this and related subjects pertaining to sexual matters.[1] That this was a relatively unexplored area of Portuguese history and legal medicine was emphasized by de Aguiar – he noted that 'Não abundam em Portugal alusões a antigos casos de hermafroditismo' (There is not an abundance of allusions in Portugal to historical cases of hermaphroditism)[2] – and despite some other studies from the 1920s and 1930s,[3] until recently this lack of attention has prevailed.[4] The state of historiography in respect of studies on hermaphroditism and sex change in Spain, by contrast, is more advanced, not only in terms of historical cultural analysis, for example on the subject of 'gender subversion' in Golden Age literature and theatre, but also in terms of specific studies on dissident figures such as Catalina de Erauso;[5] but comparative research into hermaphroditism in Iberia is in its infancy. Comparisons between Iberian realities and those of Europe more generally are also sorely lacking; most volumes on hermaphroditism in this geographical sphere confine their attention to France, Germany and Britain, with the occasional reference to 'periphery' countries such as Italy and to Greece for the ancient or mythological background.[6]

The purpose of this study is to deepen our knowledge of discussions on theories about cases of hermaphroditism and sex change in Iberia over the years 1500 to 1800. Although our principal focus will be on discourse on hermaphroditism and sex change in Spain, with only one chapter specifically on Portugal, we hope to suggest that given the nature of the institutions that were common to both countries – such as the Inquisition, the transfer of knowledge between theological and university centres such as Coimbra and Salamanca, and the very fact of migration, especially of Jews and *conversos*, between Spain and Portugal – there were commonalities across Iberia in respect of the phenomenon of hermaphroditism.[7]

The existence of the Inquisition, established in Spain in 1478 and Portugal in 1536, despite different competencies according to region, provided a set of knowledges and languages about a range of issues relating to religious observance and, not least, to matters of a sexual nature. While the question of sodomy – that 'utterly confused category' as Foucault has called it, given the range of practices that it could describe[8] – became included in the remit of the Inquisition in Aragon, it was not the case in Castile, where it was under the jurisdiction of the civil authorities;[9] in Portugal it was firmly placed in the remit of the Inquisition from the latter's inception. The life story of one scientific figure who, among a vast range of other topics, wrote about hermaphroditism, Issac Cardoso, was typical of the kind of communication and interconnections between Spanish and Portuguese medical and philosophical commentators on questions of 'natural history'. That many of these philosophers were of Jewish background is worthy of comment. Although it is not argued here that there was anything specific about Jewish thought on hermaphroditism, figures such as Cardoso were typical of the overlap between exiled individuals and scientific thought.[10] His parents were Spanish *conversos* who lived in Portugal until 1610 when they returned to Medina de Rioseco.[11] He went to Italy in 1648, settling in Venice and then Verona. In 1673 he published his *Philosophia libera* at Venice in homage to the Venetian authorities. As the historian Henry Kamen observes, 'Cardoso's story was one among many, for throughout those generations there was a constant movement of conversos between Portugal and Spain, as well as a regular migration of the exiles in Europe from one Jewish community to another'.[12] Other scientists and philosophers, such as Amatus Lusitanus, who was cited freely in Spanish sources in Latin or in the vernacular, is an example of cross-Iberian discourse on the subject of hermaphroditism itself.

In addition to this comparative approach, in contrast to some of the studies referred to above, we place particular emphasis on the medical, legal, philosophical, cultural and theological rationales and debates which framed hermaphroditism in this period. Such debates provided the framework whereby ambivalent sex was made intelligible to those who studied the subject in Iberia from the sixteenth to the nineteenth century as a product of the mentality of the times. In order not to engage in what Lucien Febvre called 'psychological anachronism',[13] it is important not only to examine in detail the thought emerging from a range of sources but also to provide a mental map whereby questions such as 'sex' and 'gender' can be successfully interrogated, without imposing current interpretations on these subjects and without necessarily adopting, for example, any strict association between, on the one hand, culture and gender and, on the other, biology and maleness or femaleness as explanatory paradigms.

The 'One-Sex' Model and 'Gender' Performativity

In order to understand the social conditions that provide a context for and temper the circulation of ideas in any given society and historical period, we need to attempt to determine the 'tacit truths' that make up a culture.[14] This is what we need to do when the question of personal identity is examined and, more specifically, when sexual identity is interrogated in early modern societies. The distinction between men and women, understood as a given in the daily life of the *Ancien Régime*, only became problematic when cracks and tensions emerged around it. During this period, some individuals were pronounced hermaphrodites, there were cases of changing sex, or people adopted the appearance of the other sex, and these are documented extensively in medical, juridical, theological and literary sources of the sixteenth and seventeenth centuries. These examples reveal clearly how society at that time dealt with the question of sexual identity.

To put this in the terms employed by Judith Butler, the construction of gendered identities takes place by means of a process of exclusion. Those figures and behaviours that do not observe, or that are excessive to, dichotomous sex and gender parameters and that operate outside of 'normal' human practice form the subjects of this kind of analysis. As these abject behaviours can only be considered from the perspective of the normative matrix that suppresses them, they cannot constitute an absolute externality to that matrix but an expression of the dynamics that occur within it.[15] To focus on Spain alone for a moment, as we have stated and despite the fact that research on these questions has not been as extensive as for some northern European countries, it is no longer possible to say that 'sexually intermediate' individuals have been without their historian. Furthermore, the role of Spanish theologians, juridical experts, doctors and literary figures in the construction of Western Christian normativity on the subject of hermaphroditism has now been shown to have been of the first order.[16]

Any examination of the practices and representations which occurred around sexually ambiguous persons in modern Spain must necessarily draw on the work of Michel Foucault and Thomas Laqueur. Laqueur, especially, has marked out the terrain of historiographical analysis on this front in the last fifteen years.[17] Both authors have not only shown that sexual dimorphism (the identification of women and men as possessing distinct and incommensurable biological differences) is a relatively recent historical phenomenon; at the same time, they have separated gender characteristics (masculine/feminine), the basis of androcentric domination, from the biological dualism of the sexes (male/female). Gender divisions do not necessarily depend intellectually on sex differentiation. The idea that the distinction between men and women is based on biological differences between the sexes would be a cultural understanding only consolidated by medical knowledge in the eighteenth century, thus breaking with the predominance of the hierarchical Hippocratic 'one-sex' model that was up to then hegemonic. This cultural understanding was also brought about (as Laqueur has argued) by the new liberal political order or by the emergence of modern biopolitics and the nineteenth-century desire to determine the 'true sex' in cases of doubtful identity (as Foucault has argued).

In this way, the work of Judith Butler is once more illuminating in respect of the characteristics and functioning of what she has called the 'heterosexual matrix'. This device functions on the basis of the performative reiteration of the norms that constitute it, a process that is never finished or ever fully stable. Its ever-changing nature produces abject practices and subjects that can become a source of subversive possibilities, and in this way, we can observe how the matrix is itself subject to a variety of changeable historical configurations that can be challenged at any given moment. Butler considers that intersexuality and trans-sexualism, excluded as categories of abjection, constitute subversive reiterations of the heterosexual matrix.[18]

It is undeniable that both Laqueur's and Foucault's work on the sexed body has supposed a form of analysis that has left behind the naturalist focus offered by the biological sciences.[19] As any other object of sociological analysis, the sexed body must be understood in the context of a strict historical analysis.[20] The task for the social scientist and historian is to reconstruct, through comparison, the series of historical configurations, structures and differences inhabited by the subject. In other words, any analysis from a social science perspective should always place the body in a precise spatial and chronological framework. Too often in the history of the biological sciences, historical sources are interrogated as if they existed atemporally, to be understood with reference to existing legal frameworks and scientific terminology.[21] Such a methodology eliminates precisely what is of interest to the historian, who comprehends the body within a specific contextual and cultural context in which the body makes and is made up by predominating practices in any given society.

Foucault's and Laqueur's work on the history of sexual dimorphism has been fundamental for the understanding of the body as the subject of sociological analysis. The work of the sociologist does not rely on the discussion of empirical propositions, that is, falsifiable affirmations such as an examination of the 'errors' made by Renaissance anatomists on questions of 'menstruating men' or sudden changes of sex. Historians and social scientists deal with performative questions, not those of 'truthfulness' or 'falsehood'; they are interested in the 'grammatical propositions' of Wittgenstein and the 'énoncés' of Foucault.[22] This distinction does not necessarily mean that it is not possible to analyse the past by drawing on current scientific terminology, but it does mean that we are explicit about our intentions and our epistemological outlook.[23]

A second obstacle that must be heeded is the tendency to treat past forms of sexual ambiguity too benevolently. This kind of retrospective utopianism is in part a result of the relativization of the present that accompanies all historical analysis of sexuality. It is an error that Foucault appears to succumb to when he considers hermaphrodites in the Middle Ages and Renaissance whereby supposedly, on reaching adulthood, they could elect the sex they wished provided that they stayed faithful to that sex for the whole of their lives.[24] In fact this practice, which was permitted by certain legal traditions, only applied to exceptional cases such as those hermaphrodites whose predominant sex on birth could not be identified. Laqueur, for his part, has detected this excessive utopianism in Foucault's work. Given that the distinctions between the genders was not founded upon any biological differences between the sexes, these distinctions, as in any patriarchal order, were based on strict prohibitions and punishments that would be applied to any transgressor.[25] In other words, the extensive understanding of hermaphrodites as naturally occurring possibilities and not monsters in human form did not in reality allow greater flexibility or rights for these subjects during the modern period in comparison to their treatment from the nineteenth century onwards. In this century, they became teratological figures whose deformities hid their 'true sex'.

Thirdly, it is important that historical research does not fall into another trap in the understanding of this phenomenon. On the one hand, the attempt to discover a hegemonic model, a monist schema or 'one-sex' model, a dualist interpretation or the visualizing of three sexes at any given time should be avoided.[26] On the other hand, it would be a mistake to avoid trying to perceive certain idealized types to aid analysis on the basis that the enormous diversity of models presented by the sources would impede such an undertaking. The first option could lead to a certain degree of dogmatism whereby the documents are only read in accordance with *a priori* understandings in an *ad hoc* manner and any evidence to the contrary safely eliminated. To some degree Laqueur has followed this route, as have some of his detractors.[27]

The inverse scenario would result in a kind of diffuse empiricism whereby models were rejected resulting in a straightforward reading of what is said in the documents. Such a stance can be seen at times in the work of Joan Cadden in her examination of the plurality of understandings current in medieval medicine.[28] Nevertheless, it is true to say, as Cadden points out, that not all references to Aristotle's theories of generation in medieval times imply the hegemony of a dualist model of the sexes, which was opposed to the recognition of sex changes or real hermaphrodites.[29] But it is important to distinguish between the purely empirical reality of quotations from Aristotelian works and the ideal type that authors such as Lorraine Daston and Katharine Park designate, as a form of abbreviation (in the same way as Max Weber wrote about the 'Protestant ethic'), of the 'Aristotelian model' whose roots can be traced in *De Generatione Animalium*.[30]

Aristotelian, Galenic and Hippocratic Interpretations and Social Rank

In addition to recognizing the shortcomings of Laqueur's approach and thereby accepting that a dualist Aristotelian model coexisted with a Hippocratic-Galenic understanding, as Lorraine Daston and Katherine Park have suggested, there is an important distinction to be made between the 'heterosexual matrix' of this period and that which emerged during the Enlightenment. In the same way as its Hippocratic counterpart, the Aristotelian interpretation was vertical and hierarchical and represented women as mutilated or accidental men.[31] This had little in common with the Enlightenment and nineteenth-century understanding of incommensurable differences between men and women situated on a horizontal plane. The Aristotelian interpretation of the sexes did not consider biology, the sex of the organism, as the stable basis whereby visible differences in culture, behaviour and roles as established by gender were in fact played out. Beyond the divide of dimorphism, Renaissance[32] and Baroque cultures gave preference to a teleological *continuum* that assimilated identities as a succession of 'folds'[33] and that established the male sex/gender as the normative frame.

Rather than the coexistence of two incompatible models (the Aristotelian and Hippocratic-Galenic), therefore, what prevailed in the modern period was a particular expression of a sex/gender regime or heterosexual matrix inhabited by a range of identities and behaviours which were in turn prone to controversy in the juridical, medical and theological fields. The challenge in this light is not to argue over which model was predominant in Spanish or Portuguese culture during the 1600s but rather to analyse the specificity and the ambivalences of this sex/gender regime within the particular time scale that is interrogated.

In this sense, we should also be careful when analysing the culture of the *Ancien Régime* not to fall into an anachronistic interpretation of the very categories of sex and gender. Following Laqueur, we know that in Europe in the sixteenth and seventeenth centuries strict differences between the social sexes, that is, in terms of what could be expressed as 'gender', were not based on biological differences.[34] To be a woman or a man was not so much to possess a particular biological quality but rather to display a social attribute. In sources recounting news of prodigies or tragic occurrences (the 'relaciones de sucesos') in Spain in the sixteenth and seventeenth centuries, it is fairly common to find mentions of sex change and the existence of hermaphrodites. In this kind of literature, sex is referred to as a 'habit' or 'state' (the famous Elena de Céspedes around 1587 spoke of 'taking the habit of a man'; Catalina de Erauso, who decided to live as a man in 1600, spoke of 'declaring her state'; and the nun Fernanda Fernández in 1792 said 'I have taken the habit of a man' in order to refer to her change of sex).[35] To be one sex or the other was like belonging to what we could designate as a particular rank or social status.[36] To change sex was to take a different kind of state, similar to that transited between singlehood and marriage. Just as one may be a noble or a vassal, one was a man or a woman. To belong to one or the other meant a series of privileges or prerogatives that the other did not have. Just as someone could not carry a sword or display certain signs of privilege, in accordance with rules governing clothing and presentation,[37] neither could a man dress as a woman or vice versa, except in exceptional circumstances such as in the theatre, masquerades or as a result of special permission granted by an ecclesiastical authority. But in the light of this, it cannot be said that in the *Ancien Régime* sex was subordinated to gender or that society somehow subscribed, *avant la lettre*, to a 'social constructionist' perspective.[38] Instead, the distinction between sex and gender was meaningless, a peculiarity of the 'heterosexual matrix' of the period.[39]

The biological in this period was never presented as the purely biological or as 'bare life'.[40] It was understood from two perspectives: on the one hand, it expressed a transcendent order, that of Nature as a 'vital force' and as a moral sphere ruled by God. *Physis* was not understood as something static; it was assimilated as part of an expression of dynamism whereby there prevailed an incessant generation of form that in turn displayed the infinite power of the *Summum Artifex*.[41] The occurrence of extraordinary or 'marvellous' events expressed, as part of a tradition that went back to Saint Agustin, the omnipotence of divine will.[42] The same occurred with the birth of hermaphrodites or with sex changes (understood as 'improvements') from women to men. At the same time, divine creation was thought to be manifest in human reproduction, and this demanded the existence of fully differentiated men and women.[43] There was no contradiction between a form of biology that allowed for intermediate figures in terms of sex and an institutional framework that excluded them in 'gender' terms. Both possibilities were inscribed in Nature, which was understood as a manifestation of divine will. For this reason, it has been argued that 'biopolitics', a power that immunizes 'bare life', could only come about in the gap left by the disappearance of the previous theologically based order.[44]

On the other hand, the second understanding of the biological, in addition to this vertical conception, linked the body and personal identity by means of a horizontal network of lineages, corporations and familiar relations. One's name, rights, obligations and prerogatives involved the body in a network of honours and dependencies. This 'deployment of alliances' implied a particular regime of visibility.[45] As such, faced with the physical presence of an unfamiliar person, it was not a question of deciphering their true self or authentic person but rather of discerning from which family or house they came. It was a case of identifying the signs that showed their rank and if they were allowed to carry these signs *de iure*.[46] This was in order to prevent fraud and cases of false identity, something that was liable to affect courtly, community and family relationships.[47] This concern was writ large at the end of the Middle Ages and beginnings of the modern period, a result of the increased mobility of populations towards the towns and cities and their inability to become inserted in the different social strata that made up these urban centres.[48] The upsurge in vagabonds, wanderers, indigents and foreigners placed outside the social networks of dependency converted these people who had no outwards signs of identity into threats to the established order.

At the same time, artistic creation – with the Baroque predilection for deception and masquerade – in the form of the short novel[49] and popular forms of theatre played with these performances (the king disguised as a beggar, the prince as a savage and the rich man as a vagabond), with the result that the social order was momentarily upset only to revert to normality at the end of the piece. In this context we can understand what Greenblatt has called 'cross-dressed theatre' with ample displays of cross-dressing and cross-coupling.[50] Having been initiated in Italy, this variety of comedy enjoyed extraordinary success in Spain in the Golden Age, where the woman disguised as a man became the dominant expression of the phenomenon rather than the reverse.[51] This 'female to male' preference has been understood as part of the teleological order whereby the dynamism of Nature would always tend towards perfection, understood as the masculine form.[52] As we will argue later, this transvested theatre would also function with a degree of ambivalence. It reinforced the sex/gender hierarchy, as everything eventually returned to its place; but at the same time, as the artificiality and the contingency of this order was exposed for all to see, it could also contribute to its subversion.

A similar experience of rupture or disturbance can be seen in those cases of hermaphrodite persons or changes of sex. In these circumstances, the determination of maleness or femaleness by the family or by those authorities implicated in the case (such as midwives, doctors, judges and bishops) did not depend on any supposed deep-down real sex but on the confirmation of rank, dress and the occupation that the individual could legitimately assume.[53] In this sense, the physical aspects of a particular body functioned as a sign of rank and not as merely biological attributes. The act, inscribed in the tradition of Roman law, of assigning the 'predominant sex' rather than the 'true sex' of the newborn hermaphrodite individual obeyed this kind of concern: to determine the rank of the person, their prerogatives and associated obligations.[54]

The body, therefore, at a time when there was still no obvious schism between popular and elite culture, was not understood as a biological reality *tout court* or as a separate sphere separating the self from the rest of the world. As the 'grotesque body' explored by Bakhtin indicates, this was an exteriorized reality, a microcosmos linked to a macrocosmos through lines of influence and preferences and dislikes.[55] It functioned as a text where divine design or the honour of lineage could be read.

The fluidity and changeability of this body, whose sex could be transformed as a consequence of an abrupt shift in the person's activity or work, was such that change of sex, bearded women and menstruating or lactating men could all occur. Such phenomena occurred as an expression of divine will whose omnipotent power was capable of moulding the individual accordingly. There was no ontological discontinuity between the physical hermaphrodite (in some sense, arguably the intersexual of today) and the social hermaphrodite (again, in some respects, the transvestite or transsexual).[56] In this sense, these cases were 'marvels' (*mirabilia*),[57] strange occurrences that certainly deviated from the normal order of Nature (*praeternaturalia*) but which could not be understood as being against Nature (*contra natura*). Indeed, as we will see, a significant section of Iberian thought in the sixteenth and seventeenth centuries understood these intermediate beings not as monstrosities but as possibilities perfectly within the broad dynamics of a prolific and diverse expression of Nature.

However, any ambiguity and mobility between the sexes was subject at the same time to severe restriction. God had wanted two sexes to exist in humans in order to guarantee procreation. The task of the civil and religious authorities was to watch over the border between the sexes and to set out the criteria whereby any one intermediary individual could participate in both sexes. These same authorities would also attempt to dissuade and to punish those who attempted to transgress those established limits by making ambiguity a *modus vivendi*. For this reason, it was not uncommon that at the time of the Counter-Reformation hermaphroditism and sex change came to be associated with sodomy as the counter-natural activity *par excellence*.[58] In these cases, what are invoked are not the marvels that exalt divine power, but the maleficence (*magicus*) inherent to these events, signs of sin or a warning of danger to come.

But any more or less natural or preternatural instance of *mirabilis* and counter-natural *magicus* is only part of what the ambiguous body could represent. A third type of experience that was doubtless less frequent nevertheless draws on the same idea of metamorphosis and sexual ambivalence. This was the *miraculus*, a supernatural intervention that departed from the normal course of Nature and presented a salvationist message. The *miraculus* could include saints who changed sex as a result of divine intervention, the final resurrection of women converted into men and, on a different level, the invocation of the androgynous as a symbol of perfection uniting two contrary tendencies.[59]

How these categories operated in the context of the medical and juridical discourse on hermaphroditism and sex change in Spain over the period from 1500 until the beginning of the eighteenth century is examined in Chapter 1. Chapter 2 examines the question of hermaphroditism in cases selected in particular from the 'New World', but with an emphasis on the traffic of ideas between Spain and the Latin American colonies. Chapter 3 focuses on the decline of the 'one-sex' model and the disruptions – and surprising reversals – that this performed in knowledge on hermaphrodites in the eighteenth century in a Spanish context. In the following chapter, Chapter 4, we shift our focus to Portugal, and through the examination of a number of different cases and the theological, medical and juridical treatises of the time, we provide a broad analysis of sex change and hermaphroditism in this country and the differences in comparison to Spain and Europe more generally. In our Conclusion, we suggest similarities and differences across Iberia with respect to historical accounts of hermaphroditism and summarize the significance of our findings for the history of hermaphroditism in Europe from the sixteenth to the eighteenth century.

1 MARVELS, MONSTERS AND PRODIGIES: HERMAPHRODITES AS NATURAL PHENOMENA IN SPAIN, 1500–1700

In this chapter, we illustrate how the notions of Nature, *mirabilia* and marvels were played out in considerations of sex, sex change and hermaphroditism in Spanish culture in the sixteenth and seventeenth centuries. Crucial to our argument is the elaboration of the working of the 'one-sex' model as understood by Laqueur and the overlapping and on occasion competing theories of Galen, Aristotle and Hippocrates in order to explain these phenomena. In the next chapter, we will provide an overview of how the notion of the 'true rank' was supplanted by the medical category of the 'true sex' in the nineteenth century.

Mirabilia

The category *mirabilia* designated extraordinary beings and events, 'marvels' that showed the omnipotence and inscrutability of divine design. This tradition stretches back to the Augustinian text *De Civitate Dei*, and was followed by Saint Isidore's *Etymologies*.[1] Portents are not necessarily counter-natural beings or isolated cases with no significance; they are natural rarities that always have their analogies in the Universe. The Universe is conceived as a dense network of relations which reveal a hidden harmony, known by God but of which humans are ignorant. Because of this ignorance, humans perceive these figures as disconcerting and horrific.

It has been argued, only partly successfully, that in the later medieval period, as a result of the social and cultural crisis that took place in the fourteenth century in the wake of plagues, massacres and famines, the belief in the harmony of the Universe was undermined. The presence of the monster was understood as evidence of the work of the devil and of great calamity to come.[2] It has also been argued that this understanding would give way, in the light of the coming of science, to a naturalist perception of the monster in the context of an emerging literature on 'marvels'. This literature stimulated devotion to pleasure and a certain curiosity towards this figure, which was presented as evidence of the benevolence of God working through a prolific and diverse expression of the natural world.[3] This image would constitute the preface to seeing the monster as an error of Nature particularly from the seventeenth century onwards. This error of Nature would be a deformity that could be explained by purely immanent causes. In this way, a process of disenchantment and rationality would be complete.[4]

But we know that things are not quite so simple. Throughout the Middle Ages there prevailed a certain division between the representation of the monster as a species from exotic lands and the idea of the monster as a warning of disaster. This dual conception would come, on the one hand, from travel literature, centred on the description of marvels (species that showed the unlimited power of the Creator) and, on the other hand, from a literature of prodigies that presented monstrous individuals, not species, as a sign of evil.[5] From the sixteenth century this distinction became less clear. Both portentous individuals and species could be understood to show the hidden harmonies of the cosmos, thus proving divine will. At the same time, the tradition that saw the monster as a punishment from Providence or a sign of disaster to come survived.[6]

Rather than this teleological account, which argues for a sceptical and disenchanting modernity, what takes place in the sixteenth and seventeenth centuries is a number of mixed understandings that exalt the monster as a marvel and at the same time fear it as a manifestation of evil. This ambivalence is clear in the case of the hermaphrodite and changes of sex, despite the fact that most medical opinions classify these as natural happenings. They are still understood to represent something out of the ordinary or rare, and they are represented as such in the Spanish case in the literature on marvels and in the 'relaciones de sucesos'.

The genre of the literature on marvels[7] maintains its presence throughout the whole of the Spanish modern period,[8] and even still has its expressions beyond the end of the seventeenth century.[9] Examples of this kind of literature, mainly penned by ecclesiastics during the period 1540 to 1677, clearly show the interest in sexually intermediate persons described as 'marvels'. The notion of *mirabilia* is maintained throughout this period in this kind of text. The apparently deformed and disordered show the limitations of human intellect and the complexity and brilliance of the order imposed by God, however inscrutable to human eyes.[10]

Hermaphrodites and episodes where masculinization occurs (in contradistinction to 'feminizing' incidents) are presented in these sources under the rubric of 'natural' events, even though they are considered to be extraordinary. Usually individual cases are referred to, although there are also, following the medieval tradition of travel literature, accounts of whole exotic peoples who have the reputation of hermaphroditism. The work of Pliny, as presented by the philosopher Calliphanes, falls into this category.[11]

The description 'natural' as used to refer to hermaphrodites, manly women or viragines[12] and masculinized females[13] is not the same as our description of such phenomena as 'biological'. In this literature, sex is identified as a variant of rank. The best example of this is offered by Antonio de Torquemada (whose work is discussed by Martín del Río and Juan de la Cerda).[14] Torquemada referred to the case of a woman from Condado de Benavente (Zamora), who was married to a poor labourer. One night the woman decided to leave her husband under the disguise of some clothes stolen from a local boy. In this way she adopted the lifestyle of a man and worked as such:

> y estando así, o que la naturaleza obrase en ella con tal pujante virtud que bastase para ello, o que la imaginación intensa de verse en el hábito de hombre tuviese tanto poder que viniese a hacer el efecto, ella se convirtió en varón, y se casó con otra mujer ... y hasta que un hombre que de antes la conocía, hallándose en el lugar de donde estaba, y viendo la semejanza que tenía con la que él le había conocido, le preguntó si por ventura era su hermano, y esta mujer, hecha varón, fiándose de él, le dijo el secreto de todo que había sucedido, rogándole con gran instancia que en ninguna manera le descubriese.

> (And so, whether Nature alone was sufficient to work with such strength and virtue or whether it was the intense power of the imagination on seeing herself in the habit of a man that had this effect, she was converted into a man and married another woman ... and until another man from before recognized her, finding her in this place, and seeing that she bore a resemblance to when he had known her, asked her if by chance she was a brother and this woman, who had become a man, told him in trust and secret everything that had happened and pleaded insistently that he should reveal nothing.)[15]

The change in clothing and occupation is what unleashes her sexual transformation, through either the action of imagination or that of Nature.[16] This same logic explains how the heroines of Golden Age drama, just by dressing as men, are capable of being bestowed with male abilities as if by magic.[17] In this case that occurred in Benavente, it is as if the abandonment of her state as a married woman and the taking on of the appearance of a man brought about her bodily transformation; it is as if the physical were an external expression of her 'state'. In addition, in the light of the uncertainty that surrounded the possession of one state or another in terms of right, duties and obligations, her act of fraud consists in taking on privileges that do not correspond to her. But her status is not reducible to either biological sex or gender, and neither is reducible to the other.[18] Gender does not provide the foundation of her sex because her 'nature' is only distinguished on the basis of her dress or occupation. It is as if sex and gender were undifferentiated, making up something else: her rank in society, something stable in itself in an ordered society but susceptible to disordering and confusion in exceptional cases.

When discussing the natural character of hermaphrodites and virilized women, all these texts follow similar formats. First of all, they present their structure of argument, like the *quaestiones* or as in examinations of conscience (Martín del Río, Antonio de Fuentelapeña); they opt for a dialogical form (Alonso de Fuentes, Antonio de Torquemada, Juan de Pineda), or they adopt a narrative description (Pedro Mexía, Juan Eusebio Nieremberg, Juan de la Cerda). They all cite ancient authorities on these matters (for example, Pliny, Ovid, Hippocrates, Phlegon, Aulus Gelius and Livy) and modern authorities (Joviano Pontano, Amato Lusitano, Fulgoso, Montaigne) who affirmed the existence of hermaphrodites and masculinized women.

In addition, these authors display an eclectic array of shared medical knowledge. This includes the Hippocratic theory of humours and generation, references to Galen and Avicenna whose accounts presented the male and female organs as identical in structure but not in position,[19] and the mention of the Aristotelian teleological principle which accounted for masculinization of women as 'improvements'. In the Aristotelian vein too was the discussion over whether the woman was a weaker form of man or not.[20]

In this mixture of positions, the Hippocratic one-sex model prevails, although accounts are much nuanced. The two accounts of 'marvels' that display most medical arguments, that of Fuentelapeña and especially that of Martín del Río, seem to have come under the influence of the French medical doctor André du Laurens (1550–1609). Du Laurens was the author of *Historia Anatomica Humani Corporis* (1593) and defended, according to Michael Stolberg,[21] the Aristotelian model of the two dichotomous sexes. The reading of this work, clearly first-hand in the case of Martín del Río in 1606 and perhaps second-hand by Fuentelapeña, sounds a cautionary note over the supposed hegemony of the one-sex model in the Spanish case.

In the case of Fuentelapeña, who, in contrast to strict Aristotelian thought, nowhere doubts the existence of true hermaphrodites, the explanation of this 'marvel'[22] draws on a clearly Hippocratic account, although it is taken from Albertus Magnus.[23] The allusion to André du Laurens ('Andrés Lorenço') is occasional and serves merely to ratify his own thesis: that women who change into men are in reality 'hidden hermaphrodites', beings that possess two natures, although this has become visible only because of excessive natural heat.[24] Strictly, then, sexual transformation is not possible. Fuentelapeña includes du Laurens (and del Río) among those who hold this opinion. He is aware that du Laurens emphasizes sexual difference ('no sólo en el modo de la situación, sino que también en el número, forma y fábrica se diferencian' (not only in respect of placement but also in number, form and make-up are they differentiated)), thus placing himself against the majority which was in favour of the Hippocratic-Galenic paradigm of the single sex. Fuentelapeña finally, however, seems to move towards an Arabigo-Galenic alternative which, without admitting true sex changes (which are understood as manifestations of the invisible sex of hidden hermaphrodites), supports the 'one-sex' model:

Y finalmente otros sienten y es lo más cierto, que aunque el instrumento sea único, puede invertirse de adentro afuera, como un guante, y que de una manera será sexo viril, y femenino de la otra, pues como sienten Galeno, Egineta, Avicena, Razes y otros muchos médicos, las mujeres tienen los mesmos vasos seminales y órganos que sirven a la generación, que los hombres.

(And finally, others believe and it is certainly true that although the instrument is identical, what is inside can be inverted, like a glove, and that what is the male sex on the one hand may be the female sex on the other. Just as Galen, Paul of Aegina, Avicenna and Rhazes argue, women possess the same seminal ducts and organs that serve generation as men.)[25]

The case of Martín del Río is different. In the second edition of his work *Disquisiciones Mágicas*, from 1612, he declares that he has read the *Historia Anatómica* by Andrés de Lorenzo in 1606.[26] This reading, Del Río avers, confirmed his own understanding that supposed masculinized women were in fact 'hermaphrodites' who possessed both sexes. As a result, the transformation of sex would be nothing more than the exteriorization of the male nature that had been hitherto hidden.[27]

However, the work by du Laurens goes further than affirming the existence of hermaphrodites. Martín del Río tells us that du Laurens effectively rejects the one-sex paradigm:

Allí expone [du Laurens] al detalle cómo es falsa la doctrina médica común acerca del varón inverso escondido en la mujer. Los órganos genitales de uno y otro sexo difieren en absoluto, no sólo por su situación, sino por su número, forma y estructura.

([du Laurens] argues in detail how false is the medical doctrine that takes the inverted male to be hidden in the body of the woman. The genital organs of one and the other are fundamentally different, not only in their position, but in their number, form and structure.)[28]

The renowned demonologist and Jesuit appears to subscribe to this dualist model, a move that allows him to reinterpret two elements associated with the one-sex model. The first of these is that the woman is a failed or weakened form of man. Secondly, he revises the teleological notion referred to above ('la naturaleza siempre tiende a lo más perfecto' (nature tends towards perfection)). With respect to the first matter, he suggests, following du Laurens, that woman is not an incomplete man as in the hierarchical and vertical system of the one-sex model but a finished organism that possesses its own structure. This structure corresponds to the function of womankind: 'fue menester que la mujer tuviese la conformación que tiene, pues de otro modo no se conservaría la especie humana' (it was necessary that woman possessed the conformation she indeed possessed because, if this were not the case, human kind could not be preserved).[29] On the question of teleology, he provides a different interpretation: 'hay que decir, pues, que la naturaleza siempre procura lo más perfecto, no porque siempre tienda a engendrar varón, sino porque cuando tiende a ello procura hacerlo lo mejor posible y lo mismo cuando se propone hacer hembra' (we can say, then, that nature always seeks the most perfect, not because it will always seek to engender a male but because when it seeks the perfect it does so in the best way and in the same way as when it tries to make a woman).[30] By means of a reading of the work of du Laurens, del Río, despite not adhering to an Aristotelian dualist model (he concedes the existence of hermaphrodites), does dissent from the Hippocratic-Galenic tradition by recognizing the irreducible singularity of the female body.

The same kind of register as that used in accounts of 'marvels' can be seen in a case of sex change that took place in 1617 in the Convento de la Coronada, in Úbeda.[31] It was mentioned in a 'relación de sucesos' published the same year. As is usual in the majority of texts of this type, it appears in the form of an epistle.[32] Within appears the letter sent from the prior of the Order of Saint Domingo of Úbeda to the abbot of San Salvador, Granada. In contrast to other cases of prodigies and apparitions, in this case there was no intention of terrorizing the reader into seeking conversion or sanctuary in the Church.[33] Like other accounts appearing in this genre of literature, this text presents the masculinization of a nun. This is presented as something extraordinary but natural, to be included within the 'miracles of nature'.[34]

The nun is question was María Muñoz. Muñoz was suspected of being a man by her sisters ('echaba mano a una espada y disparaba un arcabuz y otras cosas que hacía de hombre' (she wielded a sword and fired an arquebus and did other things like a man)). She was examined by the prioress, who declared her to be a woman. However, shortly afterward, the nun wrote to the prior of Saint Domingo of Úbeda. After interviewing her and summoning the Dominican prior from Baeza, the authorities decided to examine her: 'y hallamos ser hombre perfecto en la naturaleza de hombre y que no tenía de mujer sino un agujerillo como un piñón más arriba del lugar donde dicen que las mujeres tienen su sexo a pie del que le había salido de hombre' (and we found a man perfect in the nature of man and there was nothing of a woman apart from a small hole like a pine cone above the place where it is said women have their sex, next to where a man's sex had come out).

The text shows a certain degree of familiarity with the medical knowledge of the period. The nun's sex had been changed as a result of some strenuous work in the fields; this had meant a sudden increase in heat and the expulsion of a penis from her body. The exertion provoked 'un gran dolor entre las dos ingles' (a great pain in the crotch), which had caused a swelling. After three days, the swelling went down but 'le había salido naturaleza de hombre' (a man's nature had come out). Seven days afterwards the transformation gathers pace: 'le comenzaba a negregear el bozo y se le mudó la voz muy gruesa' (her facial hair darkened and her voice became much deeper). Although the document is not entirely clear on this point, the illustration that accompanies the case suggests that the nun was indeed a hermaphrodite.[35]

Be that as it may, the episode confirms that the text belongs to a period that did not distinguish between biological sex and gender. Such is confirmed by the attitude of the father of María Muñoz. His first reaction was of shock, but this did not last: 'el padre está muy contento, porque es hombre rico y no tenía heredero' (the father is very happy as he is a rich man and had no heir). It is clear from such statements that sex was not understood as a deep-seated biological reality. Instead it was related to kinship, blood lines, worldly goods and names that make up a lineage. Nowadays, the difficulty for the parents of the trans-sexual is in accepting without problems the psychic peculiarity of their offspring so that their mental health and self-esteem are not harmed; in the seventeenth century the masculinized woman posed the question of whether there would need to be a change in the transmission of inheritance and the family name and if it affected the rights of other descendants.

María Muñoz herself received her new sex not as a kind of physical novelty but as the promotion to a higher rank from nun to firstborn and only heir. This would afford her access to privileges previously thought impossible. This change in circumstance, although unusual in an order based on status that was not used to sudden alterations, merited being communicated to the highest temporal authority that could accord such privileges. For this reason the author of the letter recommends that the king should be written to in order to communicate the 'strangeness' of the case.

Galenisms and Aristotelianisms

Historians of Spanish medicine agree that throughout the sixteenth century Galenism was the primary medical doctrine.[36] But this was an unusual form of Galenism which was derived less from the great medieval commentaries, including Arabic texts such as Avicenna's *Canon*, than from a contrasting reading with the classics of Greek medicine.[37] The humanist return to the original texts had also been followed in the area of medicine. Galenism, enriched primarily by its contacts with Hippocratic referents,[38] and to a lesser degree through Aristotelian influences, dominated the medieval intellectual scene in medicine. In addition, the field was open to the anatomical teachings of Vesalio and other Italian masters.

This hegemony of the Galenic tradition undoubtedly led to the predominance of the one-sex model, a position that was practically uncontested throughout the second half of the sixteenth century. This does not mean, however, that Aristotle's thought on the sexes and generation was ignored in the Renaissance period in Spain. What happened was that Aristotelian understandings did not significantly undermine Galenic-Hippocratic predominance and its one-sex model.

Within this framework, women were described as having the same organs of generation as men. The difference resided in their position: they were internal to women but external to men. Another matter was that of temperament, which

was related with the positioning of these organs. The lack of heat due to the coldness and the moistness of women explains why the organs such as the 'madre',
analogous to the 'verga' and the 'testicles' or 'compañones', remained inside the
body. This internal arrangement was in order to allow conception to take place.
Doctors not only invoked Galen to prove this point but also referred to observation and illustration. Anatomists such as Bernardino de Montaña (1489–1558),[39]
Juan Valverde de Amusco (1525–88),[40] Juan Fragoso (1530–97)[41] and Andrés de
León (1560–1602),[42] theologians who wrote texts on the 'conservation of health'
such as Blas Álvarez de Miravall (fl. 1597),[43] and doctors that published influential books such as Juan Huarte de San Juan (1529–88)[44] contributed to the
consolidation of the representation of woman as an imperfect man.

The recourse to the idea of one sex corresponds to an understanding of generation that combines Hippocratic and Galenic theory. The former insisted that
generation was the result of the mixing of male semen (predominantly warm
and dry) with female semen (moist and cool) in the interior of the 'madre'.[45] This
is usually represented as formed by seven cavities. If during the mixing process
the male semen was dominant and contact occurred in the right three cavities,
a male would be born. For this reason mothers were advised to lie on their right
sides if they wished for a male child. If the female seed was victorious and the
fusion took place in the left three 'cells', a girl would be born.[46] If during the
process neither the male nor the female type of semen prevailed and the mixture occurred in the central cavity, a hermaphrodite would be born. Hippocratic
theory required the emission of female seed for fertilization to take place. It was
for this reason that doctors following this doctrine referred to the lack of sexual
appetite among sterile women.[47]

On the other hand, the Galenic doctrine, which was adopted by an important sector of Spanish doctors, argued for some differences with respect to the
Hippocratic model. Firstly, the more important role and the superior strength
of male semen in terms of its ability to fertilize over that of female semen were
emphasized.[48] In addition, it was understood that the woman, as well as giving
seed, also supplied menstrual blood which placed a crucial role in the formation of the foetus's body.[49] This understanding of female sperm as seed, although
of inferior quality, distinguished Galenic theory from that of Aristotle, which
favoured the male seed on its own as that which had the ability to fertilize.
Female seed was reduced to mere primary material.[50]

Within this framework, faithful to the one-sex model which dominated
Spanish thought in the 1500s, a huge range of intermediates was understood
to exist. Between the most perfect (the virile man) and the most imperfect (the
feminine woman), the existence of hybrid and itinerant figures was permitted.
These included hermaphrodites or androgynes, whose existence supported a
long philosophical and literary tradition and whose formation was explained

by recourse to the Hippocratic mixing of semen. Not only was their existence accepted and believed to be common in exotic lands,[51] they were also classified into several categories.[52] Indeed, their presence in Castile was also recorded.[53]

The prestigious Leonese anatomist Luis Mercado (1525–1611), who became court doctor under Felipe II, appeared to dissent from the current majority by considering hermaphrodites in the third book of his *De Mulierum Affectionibus* (1579) among the varieties of monster.[54] Following the arguments of *De Generatione Animalium*, Mercado explained the monster as a result of a disproportion between material and form (*vis formativa*) which had taken place in the generation process. The result was the disruption of the similarities between offspring and parents.[55] However, reinterpreting Aristotle on this point, Mercado did not consider the monster as a being contrary to the intentions of Nature. The monster was a 'preternatural' figure, that is, it only deviated from the order established by Providence to the extent of what was usual in that order.[56] Nowhere does he deny, in contrast with Aristotle's interpretation, the existence of true hermaphrodites, although he believes that any claims of sex change are fables. These would come about from the confusion arising from disproportionately large labia in some women or from cases of a prolapsed uterus. Finally, the beard of some women was nothing more than an indication of the suppression of menstruation.[57]

Together with the hermaphrodite, the existence of other transitory forms within the one-sex model was recognized. These included menstruating males,[58] manly women or viragines,[59] 'soft' men ('mariosos'),[60] those capable of giving birth[61] and lactating men.[62] Finally, there was frequent mention of individuals who had changed sex. While these were nearly always cases of masculinization, not all doctors excluded the possibility of feminization, although this was considered exceptional.[63] In the same way as in the literature on marvels, past and contemporary accounts were cited that gave credence to these stories. The similarities and symmetries between male and female organs and recourse to the humoral theory in order to explain alterations in the humours, as a result of an excess or lack of heat, supplied a coherent basis from which to explain cases of sexual metamorphosis. Transmutations could take place before birth, as occurred with the 'hombres mariosos' of Fragoso, Huarte de San Juan and Andrés de León,[64] or could be the result of post-natal changes. Fuentelapeña refers to these transformations, implying that men could change into women and could give birth. However, as has already been suggested, these alterations are put down to the hidden presence of the genitalia of the other sex; these feminized men would in fact be 'hidden hermaphrodites'. Pedro Bovistuau, on the other hand, differentiated between spontaneous transformation, when Nature, following its course, tended towards perfection, and those cases of voluntary change (Heliogabalus and Nero).[65]

Doctors, however, maintained almost complete unanimity on rejecting post-natal changes. This would have implied, as Paolo Zacchia noted in the seventeenth century, a form of degeneration contrary to the law of physics and against natural law which tended towards perfection.[66] As shall be seen below, such loss of masculinity also became understood within the social significance of the decline of empire and the erosion of masculinity.

From a close reading of Spanish medical texts from the seventeenth century, we can see that some of these postulates on the representation of the sexes began to change. While we cannot talk of an intellectual break as such, as the anatomical textbooks of the previous century continued to be republished and because the Galenic dominance continued practically undisturbed, we can point to some developments at the beginning of the century. These concerned the differences between the sexes, and they affected understandings of cases of hermaphroditism and transmutation of sex.

It seems to be certain that the work of André du Laurens (1558–1609), *Historia Anatomica Humani Corporis*, published for the first time in 1593, played an important role in this development. Du Laurens, of Jewish descent, professor at Montpellier and physician to Henri IV, questioned the identical content and structure of the male and female body in Galenic theory and denied the possibility of sex change. This does not mean that he subscribed in any simple manner to a two-sex dichotomous model in accordance with the Aristotelianism of Jean Riolan,[67] as Stolberg has supposed.[68] But neither does it mean that his differences with prevailing Galenic-Hippocratic hypotheses can be minimized, as Laqueur would have it.[69]

There were a number of doctors who agreed with the arguments of du Laurens. These include Pedro García Carrero (1555–1628), professor at Alcalá de Henares and court physician to Felipe III. García Carrero is considered by historians as 'la personalidad más importante de la escuela médica complutense a comienzos del siglo XVII' (the most important personality of the Complutense University's medical school at the beginning of the seventeenth century).[70] This author examined the question of generation, the anatomy of the sexes and the genesis of monsters in his extensive *Disputationes Medicae* (1605). As we noted in the case of Luis Mercado, the scholastic influence together with Aristotelianism were present in a work whose argumentative structure followed the university convention of the 'quaestio disputata'.

García Carrero discussed sexual difference but energetically denied that woman could be considered as a failure of Nature or as an imperfect male, a sort of monster.[71] He understands female anatomy to be composed with the aim of generation. The female body is not a mistake or an error, although he does admit that the female constitution is weaker and colder than that of the man. Further, he recognizes, along with Aristotle, that the female body only fulfils the role of material cause in generation.[72] García Carrero rejects, therefore, one of the elements of the monist model – the understanding of woman as an imperfect male.

In addition, García Carrero disagrees with Galen in that the difference between men and women goes beyond the mere position of the genitalia (internal in the woman, external in the male). He suggests that 'differunt enim aliter', although he does not specify precisely what these differences are. He also dedicates a number of chapters to the question of the monster and the genesis of different varieties. Here, García Carrero adheres fundamentally to the argument sustained by Aristotle in his *De Generatione Animalium*. But he differs from Aristotle in his consideration of the monster not as an error of nature ('deffectus naturae'),[73] since Nature can never be wrong, but as an entity 'praeter naturam', that is, a deviation with respect to the normal order provided by Nature. He relies on Aristotelian theory in order to explain the monster as a consequence of disequilibrium (of magnitude, quantity, etc.) between the *vis formativa* and the material used for generation. From this perspective, the hermaphrodite is classified as a monster – a creature with both sexes which can appear in human and animal species.[74]

Following the argument of *De Generatione Animalium*, hermaphrodites would be the consequence of an overabundance of material entailed by female menstrual blood. This excess would not allow for the formation of two distinct foetuses, although more material than was required for one foetus would be present. This material, where the male semen was predominant in certain parts and the female semen in others, would go on to form the genitalia of the other sex. In any case, according to Aristotle, the hermaphrodite would only be apparent since the sex of the foetus did not depend on the genitalia but on the heat of its heart.[75]

García Carrero includes the hermaphrodite among monsters, and even if he is not specific on their genesis, given his Aristotelian references it can be argued that the Spanish doctor believed that excess material was indeed the principal cause.[76] On the other hand, he does not sustain explicitly that true hermaphrodites are a fiction, and he does not allude to the determination of sex by means of 'cordial' heat. However, for García Carrero, in hermaphrodites one sex is present in atrophied form and the other active.[77] Does this mean that García Carrero denies the existence of human hermaphrodites? Rather than this, it would appear that García Carrero, under the influence of Aristotle and du Laurens, whom he also cites frequently,[78] embraces a perspective that does not tie in exactly with the one-sex model prevalent during the preceding century. This does not mean, however, that full Aristotelianism was embraced by García Carrero. Indeed, in his later *Disputationes medicae super fen primam libri primi Avicennae* (1611), he sustained, following the prevailing Hippocratic orthodoxy, that female seed took an active and necessary role in generation. He even defended the hypothesis that viragines or women with large clitorises were capable of inseminating and giving birth as a result of intercourse with another female rich in seed.[79]

The influential juridical figure Alonso Carranza (fl. 1625) trod a similar path to García Carrero. Carranza was the author of *Disputatio de vera humanu partus naturalis et legitimi designationi* (1628),[80] and he argued, like Luis de Mercado and Rodrigo de Castro,[81] that hermaphroditism was a monstrous condition. Citing André du Laurens, among others, he considered hermaphroditism to be 'peccatum in sexu',[82] that is, an 'error' of sex.[83] However, in contrast to those who considered the hermaphrodite to be a mere figment, or fiction or simulacrum, Carranza believed in their existence in reality.[84] He classified them according to four types: masculine, with a fully formed male sex but with a false vulva that distilled no liquid whatsoever; feminine, with a vulva and menstrual flow, who possessed above their vulva a sort of male member which underwent a species of erection but without testicles or scrotum, and which was incapable of producing semen; the third type possessed both sexes located in opposite places, and even though they emitted a kind of semen and were capable of urination, they could not procreate; and the fourth class consisted of those who had perfectly formed male and female genitalia and were 'potent', and who possessed on the right side a male breast and on the left a woman's breast.[85]

A further step in the critique of the one-sex model was taken by Spanish medicine during the 1600s by Gaspar Bravo de Sobremonte (1603–83). Bravo de Sobremonte was professor at the University of Valladolid, physician to the court of Felipe IV and Carlos II and the primary medical advisor for the Inquisition. He was one of the most important anatomists in Spanish medicine of the time.[86] In a 'promptuario' in the third volume of his *Opera Medicinalia*, published in 1671, and in one of the three 'disputationes' that make up the fourth volume, published in 1679, Bravo de Sobremonte discussed the issue of hermaphroditism and sex change.[87]

After surveying the different authorities on the question and a number of accounts that alleged the existence of sexual transformation (Ovid, Plato, Herodotus, Hippocrates, Paré, Montaigne, Torreblanca Villalpando, Nieremberg, del Río, etc.), and after referring to several Spanish cases (the famous Brígida de Peñaranda and metamorphoses that occurred in Madrid, Alcalá de Henares and Córdoba), Bravo de Sobremonte outlined his arguments against the reality of such events as impossible in Nature. In order to argue this point, he began by refuting the Aristotelian dictum as already seen in the work of García Carrero: Nature does not produce woman through error, as though a mistake had been made while trying to create something more perfect, i.e. the male.[88] The female is not a monster or a failed male but instead a figure engendered by Nature, one who possesses certain peculiarities allowing her to procreate and thus guarantee human survival.[89] For this reason, the female was not essentially different from the male; she was no monster. Any difference was merely accidental.[90]

In arguing for this difference, Bravo de Sobremonte attacks Galenic thought on the subject by noting that not only are the scrotum and uterus in a different place, but their conformation, quantity and magnitude also diverge. They are not, therefore, an entity as supposed by Galen.[91] Any refutation could not be clearer

– the distinction between male and female genitalia makes sex changes impossible because any excess heat in the case of the expulsion of the female organs to the exterior would not be sufficient to transform them into male organs. The difference is not just one of position; it affects the attributes of both sexes in many senses. Bravo de Sobremonte does not merely argue for this duality but also carefully compares the male and female organs (penis, uterus, testicles and scrotum) in order to back up his thesis of difference.[92] Among the authorities he refers to in order to do so are García Carrero, André du Laurens and Jean Riolan.

Any sex change or conversion of the organs of generation from one sex to the other is impossible given their incommensurable differences.[93] How then are women who apparently change into men to be explained? Bravo de Sobremonte suggests three explanations. Firstly, these women could be men who look like women and whose genitalia have not descended (what might now be termed chryptorchid males). When the genitalia do in fact descend, these men appear to change sex even though they have always been men. Secondly, there may be cases of women whose labia or clitoris (Bravo de Sobremonte does make the distinction) are so well developed that they are taken to be men. Thirdly, confusion could arise from the presence of 'hermaphroditae seu androgines' – both terms are used – who can move between the two sexes and are identified wrongly as having changed sex.[94]

Those individuals who do change sex belong to the realm of 'poetarum figmenta', literary fables or figments.[95] However, Bravo de Sobremonte does not deny the possibility of real hermaphrodites and even cites these cases as a source of error in supposed cases of changed sex. Bravo, who so firmly argued for the dichotomous duality of the sexes, does not go as far as neo-Aristotelians such as Constantino Variolo and Jean Riolan,[96] who reject completely the existence of true hermaphrodites. In the second part of the 'Promptuarium XXIV' of 1671, whose title is precisely 'De hermaphroditis', Bravo de Sobremonte clearly admits that such individuals exist. Although he recognizes that the medical authorities usually include them in the monster category, he indicates some types of hermaphrodite which have a particularly monstrous conformation.[97] He adopts the classification of 'androgini' proposed by Alonso Carranza and suggests four different types. He rejects the astrological explanation of the hermaphrodite and favours the 'imaginative' cause during coitus. He also has recourse to supposed imaginary cavities in the uterus which were present in the Hippocratic tradition. Despite this variety of causes, he seems to go for a Hippocratic explanation – the conformation of a hermaphrodite depends on the mixing of the seminal material and on the balance of forces between the male and female principles. Finally, confirming the impossibility of any masculinizing metamorphosis, Bravo insists that hermaphrodites possess the two sexes *ab initio* and that even they do not undergo any kind of transformation. What actually happens is that the male genitalia remain hidden until the individual develops and they become visible along with the female parts.[98]

Any critique of the one-sex model such as those contained in the work of García Carrero and Bravo de Sobremonte should not be taken as signs of an emerging rationalism that would finally exile fantastic figures such as the hermaphrodite from Western thought. On the contrary, these figures would continue in good health in the medical literature of the 1600s.[99] Even astrological, moral or theological explanations for the monster would continue and would inform not only miscellaneous texts on 'marvels' but medical thought as well.

An example of this is offered by the work of José Rivilla Bonet (fl. 1690), *Desvíos de la Naturaleza o Tratado del Origen de los Monstruos* (1695).[100] Rivilla was surgeon to the Viceroy of Perú and at the Royal Women's Hospital de la Caridad de Lima, which published this work. It was there that the author had the opportunity of observing, in November 1694, the birth of a bicephalous monster. This episode not only challenged anatomical knowledge to date but also posed a theological dilemma as to whether the newborn in question should be baptized once or twice. It was this conundrum that led Rivilla to write his book.

The book is divided into ten chapters.[101] The two first chapters discuss the significance and definition of the word 'monster' and its relation to a family of words that were semantically close, such as portentum, ostentum and prodigium. The following two chapters sketched out the taxonomy of monsters. Chapters five and six analysed the causes of such deviations. The seventh chapter looks specifically at the formation of bicephalous monsters. The final chapters constitute a detailed study of the monster born in Lima: its anatomy is described, and the questions of how many souls should be ascribed to the monster, and whether baptism should be performed on both heads, are discussed. The book finishes with an appendix in which Rivilla Bonet gives an overview of different cases he has come across, together with the array of surgical methods employed in each.

This text is situated on the borders of the medical treatise and the narration of curiosities or 'marvels', exemplified by the discussion of the bicephalous baby of Lima. In the tradition of other authors such as Riolan, du Laurens and Bahuin, who are cited, Rivilla rejects the Aristotelian definition of the monster. The monster should not be categorized as an error but as a being 'procreado fuera de la intención de la Naturaleza' (procreated outside of the intentions of Nature).[102] To the degree that this being is a result of the process of generation, a process ordered by Nature, it can be considered 'contra la naturaleza, mas no contra toda ella, sino contra su más frecuente caso' (against nature but not against nature in its entirety, rather it is against its most usual expression).[103] The Lima birth was a case of a child 'praeter naturam' and not 'contra naturam'.

Despite discussing extensively the problem of monsters, Rivilla says little about sex changes and hermaphrodites. The hermaphrodite is mentioned on one occasion, and this is just to remind the reader that animals that, given the characteristics of their species, present certain peculiarities such as 'entrambas naturalezas; ya hermafrodita con alternativa amba en los sexos' (having both natures; as a hermaphrodite with a dual alternative in its sex) should not be considered monsters.[104]

When discussing the genesis of monsters, Rivilla distinguishes between two types of causes. Firstly, Rivilla discusses superhuman or diabolical causes. Here he connects with an age-old tradition that includes Saint Isidore and Saint Agustin. The monster is understood as a punishment for the abominable sin of the parents or as the forewarning of a calamity to come.[105] These causes also include monsters engendered through carnal intercourse with the devil or as a result of certain astrological alignments.[106] The second type of causes covers 'inferior physical causes'. These are the main subject of the book. Rivilla is faithful to the explanation of monsters provided by the Aristotelian tradition, which is detailed, as we have seen, in the work by Luis Mercado, *De Mulierum Affectionibus*.

The distinction between superhuman causes and physical causes does not in itself imply that both sets of causes worked independently. Rivilla remarks that physical causes can operate as 'secondary causes' which 'puede también servirse Dios' (God can also employ) for purposes of punishment.[107] In this way, Rivilla goes beyond the reasoning of monsters as divine artefacts or preternatural rarities. The monster is not just an object of curiosity or repugnance but also one of terror because of its diabolical implications or its significance as a punishment for abominable acts. In this way, the monster seems to depart from the field of *mirabilia* to become part of the world of evil or *magicus*. This second category will be examined in the next section. For now, a brief summary of what has been said is provided.

Two different types of evidence where the problem of the configuration of the sexes occurs have been discussed. The first type of evidence, one of fuzzy borders and miscellaneous in Nature, is known as the literature of marvels. This first genre tends to accept the existence of intermediate forms such as the hermaphrodite as a natural occurrence and as a sign of divine omnipotence. The Hippocratic-Galenic schema of one sex and the Aristotelian principle that understands natural dynamism as teleologically driven towards perfection are both present in this literature. This corpus tends to reject the possibility of men changing into women but accepts unproblematically the reverse, understanding such metamorphoses as 'improvements'.

At the same time, the medical literature, in a more explicit and systematic manner, in general subscribes to the same monist model that depicts women as imperfect males. It understands, following Galen, that female genitalia are similar to those of the male, although they are positioned differently. In the light of this explanation, a whole range of intermediary figures is accepted (menstruating and lactating men, viragines, hermaphrodites and masculinized women). In general, both the literature on marvels and medical treatises represent these persons as creatures that do not contravene the natural order. They can be portents that provoke admiration or curiosity, preternatural rarities, or monsters that entail disgust and repugnance (as in Mercado). Finally, in both types of text, sex is thought of as rank. It is not a pure biological fact or a social construct alone. Nature, subject to Renaissance divination as an expression of divine will, can at the same time produce marvels that alter the distinction between male and female and base the continuity of masculinity and femininity on the fact of human procreation produced by the difference between man and woman since the time of Adam.[108]

The general thrust of these ideas remains strong until the end of the sixteenth century. From the beginning of the seventeenth century, however, a number of question marks start to appear. From the field of demonology and 'marvels', the texts by del Río and, to a lesser degree, by Antonio de Fuentelapeña begin to sketch out a challenge to the one-sex model, and this is picked up upon in medical thought. Two major medical authorities in Spain in the seventeenth century, Pedro García Carrero and Gaspar Bravo de Sobremonte, adduce such interpretations. The main source cited in both works that provides the source of this change is that of André du Laurens, whose *Historia Anatomica Humani Corporis*, originally published in 1593 and known in Spain from the early 1600s, questions the one-sex paradigm.

Something, then, is starting to change during the first decade of the century. But we should not be thinking in terms of a break with the old Hippocratic-Galenic one-sex model, as Stolberg argues. The most influential medical treatises of the previous century continued to be published and still enjoyed a huge degree of acceptance in medical schools. Furthermore, although the idea that woman is an imperfect man is critiqued, this does not by itself mean that social differences are understood to have a biological basis as in the medico-legal texts of the late eighteenth century. The framework of sex as rank remains intact, as does the idea of an onto-theological Nature that is inserted into a transcendent order that provides it with meaning and moral finality. In this sense, we can talk of a kind of 'sexual *Ancien Régime*', which would exist in parallel to other expressions of the old regime in terms of demography, politics and economics.

So, where precisely do the intellectual shifts in the 1600s occur? Firstly, the notion of the woman as a failed male or monster is placed in doubt. Instead, women are understood as being of the same species as the man, with accidental differences but produced deliberately by Nature. Nature has given her a number of specific qualities that allow for the human species to continue. Secondly, such a notion entailed the rejection of the Galenic principle of sameness between male and female genitalia. Now, difference was emphasized, not only in terms of position but also of magnitude, structure and number. Certain parts were signalled as specific to one sex, such as the clitoris, which have no correspondence with the male parts.

Thirdly, this argument eliminates the possibility of sex changes (see Bravo de Sobremonte, and to some degree del Río and Fuentelapeña). It also entails doubt with respect to the existence of true hermaphrodites (e.g. García Carrero, who accepts the existence of hermaphrodites but argues that only one sex actually functions in them, while the other is atrophied). These shifts do not allow for a two-sex model of incommensurable sexes and fundamental differences between men and women. Rather, they suggest certain cracks or anomalies within the one-sex model. These should not be exaggerated (Stolberg), but neither should they be minimized (Laqueur).

These anomalies will be resignified and reinterpreted in the light of the new understanding of sexual difference that is slowly formed from the eighteenth century onwards.[109] These shifts will become the demonstrative proofs of the new model. But for this to take place, life and Nature will have to free themselves from the old order. *Divinatio* must become immanent causal analysis, and at the same time life and Nature must become autonomous processes governed by their own internal logics. Only with the appearance of 'bare life' as the sphere in which social regulation occurs, where purely biological accounts allow for emerging political orders, will biological sex become the fundamental basis of sexual identity.

Magicus

The idea of *magicus* corresponds to what is supernaturally evil, where sin is present and where the intervention of Satan can be found. In this sense, sex changes and hermaphrodites often appear from the time of primitive Christianity up to the Renaissance as portents or warnings of some catastrophe to come or as the result of sins against Nature or sodomy. The hermaphrodite is understood as the creator of disorder and chaos that undermines the regulatory logic of Nature.

This state of affairs is shared by the hermaphrodite, the sex change and the monster. The etymology of monstruum (*monere*) or of portentum (*portendere*) shows how this significance came about. Their evil aura comes about through the production of disorder or apparent rebellion against the harmony of Nature set out by God. Both characteristics have a long history that goes back to pagan times: the tradition of the hermaphrodite as a bad sign (seen in the work of Cicero, Pliny and Livy) and as an error of Nature (Aristotle, Lucrecius and Ovid). This explains why hermaphrodites on birth were thrown into the sea in Greece and into the Tiber in Rome.[110] By contrast, Christianity, to judge from allusions in some of the sentences and *parerga* of the *Digesta* and in the *Institutiones* by Justinian, would tend to recognize the humanity of the hermaphrodite in the same way as it would with 'deformed' newborns.[111]

In addition, the Book of Genesis recalls that the division of the sexes occurs as a result of divine will. The existence of the hermaphrodite constitutes an act of rebellion against this state of affairs and thus establishes its proximity with the evil practice of sodomy,[112] an act that interrupted procreation. In primitive Christianity (for example in Tertulian and Clement of Alexandria), sodomy and hermaphroditism were confused.[113] In the same way that in the genesis of monsters the sin of luxury or satanic intervention was often present (e.g. the monster as a result of fornication with a beast or an incubus or succubus), the proximity of hermaphroditism with sodomy was often argued.

Such values are widespread in the European literature on prodigies of the 1500s in the context of religious reform. In this period, Catholics and Protestants both signalled the imminent arrival of the Antichrist and interpreted monsters and hermaphrodites as presages of an apocalyptic occurrence. In addition to the widely distributed texts of Rueff (*De Conceptu et Generatione Hominis*, 1554) and Lycosthenes (*Prodigiorum ac Ostentorum Chronicon*, 1557),[114] in Spain we see *El Sumario de las Maravillas y Espantables Cosas que en el Mundo han Acontecido* (1524) by Alvar Gutiérrez de Torres. Drawing on a tradition that went back to antiquity via Isidore of Seville and Saint Augustine, this author emphasized the sinful condition and the ominous aura that surrounds the monster.[115]

The first part of *El Sumario de Maravillas* examines the presence of different prodigies which are linked to particular historical events in the ancient world, in Christianity and in the Spanish realm. The second part focuses on questions pertaining to astrology. Among the many portents discussed – from monstrous births to earthquakes and hurricanes – two particularly related to the transgression of sexual boundaries are mentioned. The first, taken from Augustin's *City of God*, alludes to sex changes that took place during the mandate of Emperor Antoninus Pius. These metamorphoses announced future calamities: 'y las mujeres y gallinas que en hombres y en gallos fueron convertidos' (and the women and hens into men and cockerels were converted).[116] The second episode comes from the time of Rodrigo Díaz de Vivar, shortly before the Oath of Saint Agatha taken by Alfonso VI, and it describes the birth of a baby boy with two heads: 'y en el sexo y natura era doblado' (and in his sex and nature he was double).[117]

The same association between prodigies, sexual ambiguity and bad tidings was present in the context of the famous monstrous birth at Ravenna in March 1512.[118] This was an aberrant birth in many senses – one of them was its condition as a hermaphrodite. Eighteen days after the birth, a coalition of papal troops from France and Spain took Ravenna and sacked it. The news of this monster spread rapidly throughout Europe; it was commented upon in Spain too, and it left a long-lasting mark. Nearly ninety years afterwards, Mateo Alemán discussed the significance of monsters in the following terms: 'De aquestas monstruosidades tenían todos muy grande admiración; y considerando personas muy doctas que siempre semejantes monstruos suelen ser prodigiosos, pusiéronse a especular su significación' (Of all these monsters, everyone had grand admiration; and given the fact that learned people always considered them to be prodigies, their significance was speculated upon). Among the many messages that were provided by monsters through means of the extra parts the body may have had, Mateo Alemán identified the horn as signifying pride and ambition, a lack of arms was translated into an ability to perform good works, and an eye on the knee was deemed to convey vanity. The presence of two sexes signified 'sodomía y bestial bruteza' (sodomy and brute force).[119]

This fragment mentions the hermaphrodite as a monster which brought bad tidings and as a punishment for sodomy. Similar characteristics are present in an event from 1688, when news of a birth in Madrid detailed a monstrous baby that possessed both 'natures'. The boy nature was placed in the middle of the face, while the female nature was in its habitual place; there were no eyes, and the nose was absent.[120] Although the author did not dare to delve into the mysterious realm of 'cosas reservadas al Altísimo' (things that were reserved to the Almighty), he suggested that this birth could well have been a punishment for the practising of acts against Nature. In this way, the birth would constitute a moral warning for all Christians.[121] It has also been said that the monstrous birth may have had political significance, a warning about the concern growing under the reign of the weak and sickly Carlos II as the Spanish Empire waned. The hermaphrodite announced the death of the sovereign and the fall of the Empire.[122] The allusion to the hermaphrodite as a political metaphor of civil war or of the weakness or effeminacy of the monarchy (James I of England, Henry III of France) was common in Europe in the sixteen and seventeenth centuries.[123] In Spain, as we will see, in the context of imperial decline, this motif would be especially strong.

The same kinds of concern about sin *contra natura* are expressed in refrains, art and the abundant iconography of the sixteenth and seventeenth centuries (in addition to the medical literature) on the subject of bearded women.[124] These women were understood to be manly women of warm complexion and thus capable of turning nutritional material into hair, just as men did. These would be viragines, luxurious women who gave themselves over to sin. They fell into the category of 'anti-virgens', presented as creatures who were 'siniestras y de mal agüero' (sinister and of bad omen).[125]

As has already been suggested, the link between sodomy and sexual ambiguity in medical texts was very frequent (see Fragoso, Huarte de San Juan and Andrés de León). Fictional works also affirm this linkage. Included in this genre is the novel by Francisco de Lugo y Dávila, whose title was precisely *El andrógino* (1622).[126] The proclivity of 'hombres mariosos' or 'mariones' towards the nefarious crime developed from a physiological cause. These were individuals who had been born as girls but who had suffered an excess of heat in the womb and therefore had changed into men. They conserved certain traits from their female past, such as the inclination towards counter-natural sin. In this way, the thesis of the 'effeminate sodomite' not having been born at the end of the seventeenth century but, at least in the Hispanic world, a century earlier has some weight.[127]

By associating sexual ambiguity with sodomy and moving to establish holy matrimony in all social spheres, the civil and ecclesiastical authorities at the height of the Counter-Reformation would severely punish any transgression at the borderlands of the sexes. Such an association made all intermediary figures possible victims of the death penalty. An examination of the controls set in place with respect to the limits between the sexes in the sixteenth and seventeenth centuries shows how such prohibitions against hermaphrodites operated.

Any kind of retrospective utopian designs must be rejected here. Hermaphrodites at this time did not live in some kind of Arcadia whereby they could elect the sex of their choice. This idea, fuelled by certain rather free interpretations of Foucault's work, should be rejected.[128] The legislation operative in the Hispanic world, at least from the *Partidas* of Alfonso X onwards, drew on the long history of Roman law. Justinian had inserted in the *Digest* a commentary by Ulpian on the law *De statu homini* (I.5.10), which concerned persons with two sets of genitalia. In this gloss, it was suggested that the hermaphrodite should be considered to belong to the sex that predominated. Here, the celebrated motif of the *praevalet* made its appearance.[129] The law presupposed a taxonomy of hermaphrodites that was derived from Greek medicine and in particular from Paul of Aegina, a system that distinguished between male and female hermaphrodites.[130] This is the understanding that is found in the sixth Alphonsine *partida*.[131]

This desire to determine the predominant sex was connected, as we have suggested, not with the attempt to locate the 'true sex' of the ambiguous individual but rather with the need to determine the status or rank (as man or woman) in order to confer certain rights on the person. It is therefore logical that the Alphonsine norm appears in the section pertaining to wills and testaments. Women, in effect, were not able to act as witnesses for wills, could not become priests and were not able to fulfil certain roles, such as that of university professors. The problem became particularly acute in respect of questions of inheritance. Could a hermaphrodite inherit his/her parents' estate? The notary Hermenegildo de Rojas dealt precisely with this problem in his *Tractatus posthumus de incompatibilitate regnorum ac maioratuum* (1664). In principle, those hermaphrodites in whom masculinity was dominant, he argued, should have all their rights recognized. In the case of a *homo mixtus* or a predominantly female hermaphrodite, succession was not possible.[132]

The doctrine based on the notion of *praevalet* was rooted in Justinian thought. During the fourteenth century this norm was revised extensively by theologians and legal experts, and the Perugian Baldo degli Ubaldi (1327–1400) introduced substantial changes. What would take place in those cases where the predominant sex could not be determined? What drove this question was the appearance in the classification of hermaphrodites of a new category: that of the neutral hermaphrodite (with no predominant sex) and the perfect hermaphrodite (with the complete genitalia of the two sexes). In contrast with the thought of Aristotle, and under the influence of Avicenna's *Canon*,[133] many medical figures finally accepted the existence of these two new categories of hermaphrodite.

By means of the thought of Baldo degli Ubaldi, the possibility of the hermaphrodite choosing his or her own sex, the *electio* in contrast to *praevalet*, was admitted. Baldo thereby marked his differences with those observers who believed that it was necessary to assign to the individual the male category, which was the most worthy sex. It was also the sex that afforded the greatest social advantage. In any case, the person was constrained to maintain this sex and swore on oath to uphold this identity.[134]

The introduction of *iusiurandam*[135] together with the clauses on *praevalet* and *electio* were systematized by the Jesuit Tomás Sánchez (1550–1610) in *disputatio* 106 of his monumental *De sancto matrimonii sacramentum disputationum* (1601–5). In subsequent editions of this work, the autor insisted that any oath to this effect should take place before a bishop rather than the parish priest.[136]

The law, then, attempted to harmonize two simultaneous consequences derived from the all-powerful divine will: hermaphrodites were part of Nature, but their double sexual nature should be subordinated to the imperative to procreate. What is also sought in such accounts is the impeding of any sodomitical act; the hermaphrodite must not engage in sexual relations with both sexes.[137] In such cases, what destiny awaited the hermaphrodite that did not obey this imperative and moved between one sex and the other? Tomás Sánchez, in the same way as Juan de Gutiérrez in his *Quaestiones tam ad sponsalia de futuro quam matrimonia* (1618), the Augustine Ponce de León in his *De sacramento matrimoniis* (1624) or the legal expert Alonso Carranza in his *Tractatus juridicus et practicus de partu* (1629), would speak of 'culpae mortales reum' in order to refer to this figure.[138] By this, Sánchez suggested that the transgression and its punishment would be confined to the internal sphere of the conscience. As Carranza argued, those who broke this oath would receive the punishment of the Eternal Father and no physical punishment on earth.[139]

Despite these interpretations by figures such as Sánchez, the finalized legal documents would stipulate that death was the punishment for those who had sinned and committed sodomy. The death penalty for these acts would be introduced somewhat later by means of a particular interpretation of certain Roman juridical norms that the humanist Giovanni Pierio Valeriano (1477–1558) discussed in his *Hieroglyphica* (1556). In Spain, this interpretation can be found in *De sancto matrimonii sacramento* (1646), a volume written by the Jesuit Martín Pérez.

The set of rules governing hermaphrodites in the Catholic world was very different from those pertaining in Protestant countries. These differences were apparent in two main aspects of the procedure on hermaphrodites. In Lutheran and Calvinist areas, well into the seventeenth century the right of the hermaphrodite to choose his/her sex and to swear to uphold it was not recognized. In general, Protestant theologians were suspicious of what they understood as the prurient nature of Catholic, and especially Jesuit, interest in sexual matters. Catholics' discussions of these questions, in the eyes of Protestants, incited a kind of *incentivum libidinum* that overshadowed any honest Christian inquiry into discipline and knowledge.[140] The question of the predominant sex in hermaphrodites became a matter for the authorities as an 'act of the State', along the lines of what Bourdieu has suggested, drawing on the authority of medical science.[141] The dispositions adopted in Freiburg in 1610 and 1635 show precisely this development.[142]

In Catholic areas, however, the certification of predominant sex did not take place as a result of a medical examination. The decision fell to the parents of the individual, or indeed to the individual him- or herself. Advice from doctors or midwives could be taken into account in this process.[143] In the Hispanic world this drew on a long-standing tradition that went back to the fourteenth century whereby doctors could act as private and public consultants in respect of legal sentences.[144] This back-staging of doctors was what Paolo Zacchia (1584–1659) attempted to remedy in his recommendations that appeared in his *Quaestiones Medico-Legales* (1621–51). In this work he attacked the 'ignorance' of midwives and defended the male doctor's absolute authority in matters of this variety.[145] Evidently, this strategy formed part of a broader process of the undermining of female expertise in respect of the body and its replacement by the hegemonic knowledge as professed by male doctors.[146]

In addition to shifting the balance of power towards male practitioners, Zacchia was also opposed to the Catholic tradition as inaugurated under Juan de Torquemada (1388–1468), according to which the final decision on annulment for marriages for physical reasons did not ultimately depend on a medical examination. In his commentary on Gratian's decrees, Torquemada argued in favour of certain criteria that would allow for the distinguishing of the male from the female.[147] Following on from his lead, the theologians Martín Azpilcueta, Tomás Sánchez, Gaspar Hurtado and Martín Bonancina argued that the greatest authority on the predominant sexual preferences of the hermaphrodite was the hermaphrodite him- or herself.[148] Any determination thus became an act of conscience whereby medical authority was not decisive.[149]

The juridical understandings of Catholic theologians and commentators, which drew on a set of doctrines (the doctrine of *praevalet, electio, iusurandum* and capital punishment for cases of perjury or deception in respect of the chosen sex), nevertheless were not cast-iron. Their very basis, which recognized that God produced creatures with two sexes and proscribed non-procreative sexual relations, opened the door to other interpretations and to other glosses on the differences between the sexes and the heterosexual norm. This is the case of the remarkable and highly documented work of the Valencian jurist Lorenzo Matheu i Sanz (1618–80).[150]

Given the circumstances of Matheu i Sanz, son of a *contador del reino* and of the daughter of the Señera y La Llosa family, he clearly held a powerful position in legal circles. He was a lawyer of the Valencian Royal Academy by 1647, an official of the Sala and Corte in Madrid since 1659, and president of the Sala de Alcaldes in 1668, and the culmination of his career was as a regent of the Council of Aragon until his death in 1671.[151]

His *Tractatus de re criminali*, first published in Lyon in 1676, was an extensively detailed legal volume that sought to provide orientation for the practice of penal law in the Hispanic world at the time. It was widely disseminated from the date of its publication.[152] In the section devoted to legal controversies, the forty-eighth question took up the topic of hermaphrodites. This was not a new issue for the author; in 1643, while acting as a magistrate in Valencia, he conducted the trial of a supposed hermaphrodite, a Valencian clothes maker who had been brought up as a woman and who was accused of not respecting the law on the election of the predominant sex. Considered as a perfect hermaphrodite, the draper had alternated sex and had employed both sets of genitalia.

The sentence passed by Matheu i Sanz saved the accused from the death penalty. The election of sex and the oath taken by this individual had taken place in 1640 on being accused of the rape of a servant girl. He had been examined by a surgeon and declared to be a hermaphrodite with no predominant sex, but these findings were declared to be unfounded. Even though the election of sex had taken place before a court, the court had not ratified this decision with the bishop of the locality. It was therefore void. As a result of this error in the legal process, capital punishment was not applied and the accused was expelled from the city.[153]

In addition to recounting this story, Matheu i Sanz displayed his knowledge of jurisprudence on the subject of hermaphrodites and his depth of erudition on the subject. He believed that Aristotle's refusal to accept the existence of perfect hermaphrodites effectively placed limits on the all-powerful nature of *physis* – that is, the infinite power of God himself – to create beings that were beyond what was finite human knowledge and imagination.[154]

The reliance on Nature in order to resolve the controversy over hermaphrodites had already been explored by theologians such as Ponce de León (1624) and Gaspar Hurtado (1635). Both argued, in contrast to Tomás Sánchez, that the imperfect hermaphrodite, whether female or male, could marry in accordance even with his/her non-predominant sex provided that procreation was possible. The natural ability to procreate was what counted in such cases.[155]

Matheu i Sanz expanded upon this argument. Double genitalia and reproductive capacity were natural gifts that God had bestowed upon perfect hermaphrodites.[156] Marriage between two perfect hermaphrodites was completely licit;[157] both partners could also – and this is where Matheu produces a surprising and daring reading – make use of their two 'natures' as these were God-given.

By adopting this interpretation, Matheu i Sanz went against the dominant approach taken by the canons and theologians of his time, from Tomás Sánchez (1607) to Martín Pérez (1676), and from Torreblanca Villalpando (1635) to Pérez de Lara (1672).[158] In addition, by dissenting from the resolution of a case discussed by Tomás Sánchez, Matheu considered it licit for a perfect hermaphrodite, in the case of widowhood, to remarry and, if desired, to choose the other sex in these new marital arrangements.[159]

In contrast to the Aristotelian tradition, Matheu i Sanz subscribed to the legacy of neo-Platonic and Augustinian thought on the subject.[160] He believed that the myth of twin sexual natures created by Plato's *Symposium* illustrated a latent truth: that perfect hermaphrodites were a possibility in humans.[161] At the same time, and by following an interpretative framework set forth by Pérez de Moya in his *Philosophia Secreta* (1585),[162] Matheu considered that the myth narrated by Ovid on Tiresias's metamorphosis from one sex to the other in fact was an allusion to the figure of the 'hidden hermaphrodite' and the moment when this hidden sex was made visible.[163] This exegesis of ancient mythology was complemented by the reference to the hermaphrodites found in African lands as referred to by Pliny.[164]

But Matheu i Sanz did not stop at his legitimization of the dual sexual activity of perfect hermaphrodites and marriage between them. He also vigorously opposed the death sentence for those hermaphrodites that had changed sex after electing one, either male or female, as exclusive. At the height of the philological critique of sources that fanned the controversies of the Counter-Reformation,[165] Matheu i Sanz argued that the sentencing of capital punishment for wayward hermaphrodites was based on a false interpretation undertaken by Valeriani in the mid-sixteenth century on the basis of some questionable Latin sources.[166] He attacked the basis on which such arguments were made and the errors derived from them in order to countenance the death penalty and set his face against the arguments evinced by the great majority of religious commentators, that is, that the hermaphrodite who used his/her two natures was committing a juridical aberration; it was not possible to use both the active male parts and the passive female parts simultaneously or concurrently.[167] For these clerics, it was as if the subject was trying to exert contradictory rights – like trying to be an abbot and an archpriest at the same time.[168]

In Matheu's view, women by nature were passive in the process of their insemination but were active in the process of gestation.[169] Any hermaphrodite's 'bisexual' behaviour therefore had nothing to do with any revolt *contra natura* or with the sin of sodomy.[170] No penetration of the incorrect vessel was perpetrated, as perfect hermaphrodites possessed both natures. The hermaphrodite who had gone back on his/her chosen sex, deceiving the courts, also had nothing to do with the sodomite.[171] For this reason, his/her activity was to be judged by one's own conscience, and the individual would be judged on death by God.

Miraculus

Despite what we have said, it would not be correct to argue that any 'sexual nomadism' in *Ancien Régime* Spain navigated only between events deemed marvellous or 'preternatural' on the one hand, and condemnation and horror as counter-natural sins on the other. There was a kind of marginal space between these extremes occupied by the hermaphrodite or sex change as *miraculus*, as a sign of redemption. This position could be understood in a literal sense – the providential intervention of God in order to produce a sexual metamorphosis – or in an allegorical sense whereby the androgyne was understood as a symbol of original perfection or lack of difference, an emblem of the harmonious fusion of the opposites.[172] In both cases, medieval and Renaissance thought could rely on a certain reading on the classics, whether of Ovid or Plato. An example of cases in the literal sense would be instances of divine intervention in order to save female saints from rape by having them sprout beards or by converting suddenly into men.[173] The second scenario, in an allegorical sense, found its home in a mystical and esoteric tradition that starts with the revelation of Hermes Trismegistus and the Gnostics (the androgyny of Adam before the Fall, and androgyny as a symbol of Christ). Such connections only survived in the Spain of the Counter-Reformation in the ambit of alchemy.

Sexual transformation could also be the result of an act of Providence that was also loaded with portentous and salvationist significance. This includes the Christian tradition associated with a line of saints such as the Portuguese Saint Wilgefortis or Uncumber and the Spanish Saint Paula de Ávila.[174] These saints grew beards when about to be possessed by men who were to violate their exclusive devotion to Christ.[175] The transformation in these cases came about as a result of the petition of the woman, who pleaded with God to change and shake off her tormentors.[176]

The cult of these virilized women was alive and well in the fifteenth and sixteenth centuries, falling foul of the devotion of the Counter-Reformation.[177] The Castilian Paula,[178] a villager from Cardeñosa, two leagues from Ávila, would trek almost daily to the hermitage of San Segundo, patron saint of the town, in order to pray. She was accosted by a man who wished to force himself on her. To this unwanted advance, the woman responded 'in virile manner' and repulsed him. On another occasion, while walking outside of the town she was once more followed by the man, who had been hunting and had sought refuge in the rural hermitage of San Lorenzo. Saint Paula then asked God 'la diese alguna fealdad en el rostro' (to bestow some form of ugliness on her face) and thick hairs sprouted from her chin, thus chasing off her assailant.

This episode, related in various versions up to the end of the seventeenth century, was beset by contrasts: the difference between being born in a village and a town and a humble peasant birth and the elevated status of the *caballero*. It has been pointed out that most versions focus not on the virginity of the woman but on her tenacious virility.[179] It has also been suggested that the determination of the Counter-Reformation to reinforce boundaries (heretic/Catholic, Old and New Christian, marriage/cohabitation, nature/counter-nature, man/woman) would explain the decline of the cult of Santa Paula after the implementation of the Tridentine decrees. However, the 'improvement' in sex that was miraculously brought about in the damsel of Cardeñosa implied too great a relaxation of the borders between the sexes. The success of the cult of Santa Águeda, a clearly feminine figure whose breasts had been cut off through martyrdom, would replace to a large degree the bearded Paula. The first testimonies that indicate such a change come from 1595, when in the San Lorenzo hermitage, partly dedicated to Santa Paula, an image of Santa Águeda was accorded a predominant position.[180]

It would be tempting to relate the reluctance of the Counter-Reformation towards sexual transmutation to the questioning of the same kind of metamorphosis in the natural world shown in medical and demonological texts of the seventeenth century. These texts, as we have seen, were not well disposed towards sexual transformation and defended sexual difference between men and women. But there were also severe legal cautions on the subject of hermaphroditism in the sixteenth century, which often associated sexual ambiguity with sodomy. Any contrast between a supposedly tolerant pre-Tridentine sixteenth century and the mentality of a punitive seventeenth century with respect to sexual ambiguity is hard to uphold.

A different case of miraculous metamorphosis, in this case from the popular repertory of the 'cantares de ciego' from the sixteenth and seventeenth centuries, can be seen in the *Casamiento entre dos Damas*.[181] Here, the beautiful Princesa Doña Gertrudis, of the Imperial Court of Vienna, was to receive a *billet doux* from a pretender. Because of an outstanding matter with two no-gooders, the lover failed to turn up for the date and fled the city. Doña Gertrudis decides to go after him and crosses several countries, ending up in the Hellenic lands. Here she passed as a student and dressed as a man. Under the name Carlos, she manages to find a post as secretary to the Prince of the locality. The Prince's daughter, Palas, was an outstanding beauty who fell quickly in love with Carlos. The father, duly pleased, gave his consent for the marriage, and merriment was had by all. On his nuptial night Carlos told Palas that he was in fact a woman. Both agreed to keep such a revelation secret, and in this manner they spent four happy years of marriage. Given the number of rumours that began to circulate as to the true sex of Carlos, he underwent several examinations and passed all. Under these circumstances Carlos and Palas decide to offer the Virgin a novena in the hope of finding a solution. Finally, journeying over a mountain, Carlos/Gertrudis spied a unicorn which knocked her over. On her breast there formed a cross, and in turn she was transformed into a man.[182]

The contact with the unicorn sent by God is what brought about her change of sex. This animal, of potent medieval significance and often associated with luxuriousness, in the modern age underwent an important transformation. The unicorn became a symbol of Christ. Despite this, in the case of Carlos and Palas, the unicorn maintained the connotation of fecundity. God's intervention was not in order to defend the virginity of the woman Gertrudis but to allow her conversion into a suitable husband in order to continue the lineage of the Prince. Rather than a myth of purity as in the hagiographies of saints, what we have here is the myth of fertility. The intervention of Providence not only counters a virtually sacrilegious myth but also restores the continuation of the bloodline.

In reality, the message of the 'romance de ciego' is the same as that contained in the comedies and the novels of the Siglo de Oro. It was a matter of playing with an attempted fraud in status – a king dressed as a beggar, a noble in the dress of a villain, a woman disguised as a man[183] – that upset the established order. The latter, however, at the end of the story, would be restored in triumphal and gracious manner.[184] The confusion arising in *Casamiento entre dos Damas* does not just concern sex (Gertrudis appearing to be Carlos) but also social status: a princess disguised as a student and then as a servant. In the end, thanks to a miracle, the problem that drives the whole story, that of the search for ideal marriage, is happily resolved. Those of high status are married with the same, and man is united to woman in order to raise offspring that will guarantee the future of the line and the consolidation of alliances.

The miraculous conversion of women into men could also transcend individual accounts in order to be raised to the status of a theological allegory. On the final resurrection all fortunate bodies will gain the status of glory, that is, with no hint of imperfection. This understanding, growing out of the one-sex model, meant that women, as incomplete men, would finally be transformed into men. A long theological tradition underlined this happening. In the apocryphal Book of Thomas, dated from the first or second century, the Virgin Mary was transformed into a man by Jesus in order to find salvation. In the story of the martyrdom suffered by Perpetua and Felicitas, a third-century text recorded in the *Acts of the Christian Martyrs*, the first becomes a male and thus accedes to Paradise.[185] The same possibility is present in the thought of Francesc de Eiximenis in his *Lo Llibre de les Dones*, translated into Castilian Spanish in 1542 (the original Catalan edition is from 1495), whereby the woman, going from imperfection to perfection, may lose her form and become a man: 'las mugeres perderán su forma, [y] serán restituidas a la mayor dignidad y nobleza de la especie humana, que es la de varón' (women will lose their form and will be restored to the greater dignity and nobleness of the human species, which is that of the man).[186]

The collective metamorphosis evoked by the theologian in a soteriological context allows us to make one more reference to sexual ambiguity in the sphere of the miraculous: the use of the androgynous as a symbol of reconciliation of opposing values by alchemy. As is well known, the division of the masculine from the feminine traverses the whole of alchemical thought. Alchemy is not merely an intellectual stance on the world; it involves the art of initiation, a complex 'technology of the self' in which uses of materials and elements are at the same time steps on the way towards spiritual ascension whereby the follower unites his or her soul with God, in what alchemists called the 'Great Work'.[187]

In this universe of meanings dominated by the difference between man and woman (and between sulphur and mercury, sun and moon and celestial woman and the father), the image of the androgyne represents the unity of opposites, the *coincidentia oppositorum*.[188] It represents the primordial sameness of material in its equivalence to gold, which results from the fusion between sulphur and mercury, the symbol of universal transmutation of metals. In the life of the initiate, it represents the unity of the soul with God. This soteriological value enshrined in androgyny affects not only individual liberation but also the emancipation of the whole of humanity. Humanity is represented as prostrate, chained to immediacy and therefore to that which is fragmentary and passing. The return to God, to primordial unity, is achieved through the rejection of immediate experience in order to gain superior wisdom. This is what alchemy offers: a kind of *gnosis* in the strict sense of the word, which demands the death of the opposites and their fusion or marriage in order to give way to the noble, innocent soul represented by the androgyne.[189]

In this sense, the Primordial Man who represents the overcoming of sexual duality in the tradition of alchemy has much in common with Gnosticism.[190] Gnosticism values the androgynous condition of Adam, as in the apocryphal Book of Thomas, and that of Christ.[191] The distinction between the sexes would therefore be a consequence of sin; Christ, anticipating an *eschaton* that would abolish all difference, would perish as a male and be resurrected as an androgyne. As has been observed, this tradition does not end with the Gnostics – it continues in the Middle Ages in the work of authors such as John Scotus Erigena,[192] and is revived among the neo-Platonists (Marsilio Ficino)[193] and the theosophists (Jakob Böhme) of the Renaissance.[194] Alchemy takes part in this revival, as Paracelsus's work shows,[195] and Spain is not absent from this phenomenon. Spain, after the brilliant era of Llull, undergoes a period of intense renovation in the mid-sixteenth century. Felipe II was an enthusiast of this trend, and the work of Ricardo Estanihurst, Caravantes and the Valencian Luis de Centellas shows how far this doctrine was alive.[196] De Centellas refers explicitly to the 'marriage' between the 'dama que mora en el cielo' (lady that lives in the sky) and the 'otra cosa' (other thing),[197] whose fusion would give rise to the 'hijo más noble y singular' (most noble and unusual offspring), in the mode of the original androgyne.[198]

In this chapter, we have tried to capture the expressions that sexual ambiguity embodied in what we have called the sexual *Ancien Régime* in respect of the triple cultural experience of the marvellous-preternatural, the magical-maleficent and the miraculous-redemptory models.

2 SEXUAL TRANSGRESSION AND HERMAPHRODITISM: THE 'NEW WORLD' AND IMPERIAL SUBJECTIVITY

For the period 1530–1688 testimonies corresponding to approximately twenty cases of hermaphrodites and masculinized women, mainly from Castile, have been collected,[1] although some doubt has been cast on the occasional case (such as María/Magdalena Muñoz).[2] This set also includes some cases of 'sex change', that is, of women who decided to dress like a man and live as one without undergoing any process of physical change.[3] In our time, hermaphrodites and changes in biological sex belong to an ontological register that is completely different from simple transvestism or changing of appearance. This difference was not operational in what might be termed the 'old sexual regime'.[4] During that time, to have one sex was to belong to a state or a rank; biological attributes formed part of that rank, as did one's dress or the kind of occupation to which one devoted oneself.

Although the set of examples referred to above is only a proportion of any number that may exist, what is significant is that at least five of the twenty-one cases correspond to women who took religious vows.[5] In addition, we know that at least three of these – Catalina de Erauso, Elena de Céspedes and Estebanía de Valdaracete – spent some time in the army or had something to do with the carrying of arms. Life in a convent also meant that double or alternate sexual identities were possible as such an existence implied a withdrawal from the world and from conjugal exigencies. Something similar occurred in the militias.[6] The 'woman soldier', in addition to being a prominent trope in literature, was also a device often utilized by women to convert into men and hence to improve their social situation.[7]

We will examine four of these cases in detail. They have been selected because they cover the whole chronological period and, in three of the cases, because they are the most well-documented examples. These examples are: Estebanía (born in Valdaracete in 1496), Elena de Céspedes (born in Alhama, Granada, 1546), Catalina de Erauso (born in San Sebastián, 1592) and Juan Díaz Donoso (a cleric from Zafra, tried in 1634). The first of these women is included here not only because of the early date of the case but also because she has hardly been mentioned in histories to date.[8]

Estebanía de Valdaracete and the Courtly Transformation of the Spanish Nobility

The source of the Estebanía case is the *Relaciones Topográficas de Felipe II*. These accounts, begun around 1575, consist of a detailed and well-structured set of information derived from a questionnaire drawn up in each village of Spain. Even though the material was eventually only gathered from some districts, the material was sent to the king's secretary for compilation. The *Relaciones* are the first example of an *enquête*[9] to take place in Spain on a grand scale and constitute a technique of power-knowledge by which the monarch attempted to garner a mass of information about the resources of his realm. The questionnaire covered a wide range of matters for the attention of the local authorities: the geographical and administrative characteristics of the town, the quality of its land, pastures and fields, head of cattle, mines, castles, privileges granted, important buildings, religious orders, the most well-known noble lineages, etc. Among the many questions asked, number 44 sought information on 'todas las cosas notables y dignas de saberse, que fuesen a propósito para la historia y descripción de cada pueblo' (all those things notable and worthy of knowledge, which were relevant for the history and description of each village).[10]

In November 1580 the local authorities of the town of Valdaracete (Madrid) wrote up and sent their contribution to Felipe II. In point 44, after mentioning disputes with other localities, the birth of Estebanía in 1496 was recorded. The notice was mentioned under the rubric of 'casos notables y dignos de saberse', that is, it belonged to the order of the extraordinary – almost a portent, but without any negative connotations. This birth was treated as a positive event, a 'marvel' just like those mentioned in the contemporaneous books by Antonio de Torquemada and Juan de la Cerda.

A brief account is given of the life of Estebanía. At the age of twenty she was renowned for her physical strength: 'era tan suelta e tan ligera e de tan buenas fuerzas que corría y saltaba e tiraba la barra e jugaba a la pelota con tanta presteza e envoltura que en su tiempo ningún mancebo la igualaba' (she was so lithe and light and of such strength that she ran and jumped, beat others running and played ball with such skill and spirit that in her time no youth matched her).[11] Estebanía undertook what were understood to be typically masculine tasks, although her appearance did not suggest any sex different from the one ascribed at birth: 'era cosa notable de ver a la dicha correr sueltos sus cabellos largos e rubios en gran manera' (it was a notable thing to see her run around, her long fair hair loose and flowing). After leaving her town and moving around the area and becoming known for her brave deeds – probably with the use of arms as these deeds are called 'cosas tan heroicas' (such heroic deeds) – she went to Granada, where she was asked to present herself to the authorities of the Chancillería. The authorities did not accept that a woman could 'hacer cosas tan heroicas', and for this reason she was examined by 'matronas y parteras para ver su participación del sexo viril, y fue hallada ser hermafrodita' (matrons and midwives to ascertain her participation in the virile sex, and she was found to be a hermaphrodite).[12] Her examiners (no medical doctor was present in accordance with the tradition whereby females were examined by women) were not able to establish the 'predominant sex', one of the keys at the time to confirm a 'doubtful' case as either a man or a woman. Despite what we noted on the subject in the previous chapter, it would seem that the authority of male doctors had still not supplanted the role of midwives in the early decades of the sixteenth century.[13] In addition, this recourse to the criteria provided by experts, without taking into account the voice of the person affected, would appear to distance these magistrates from the recommendations provided by Martín Azpilcueta, Tomás Sánchez and other Catholic theologians who tended to consider the *praevalet* as a question that affected the conscience of the hermaphrodite or that of his/her parents.[14]

In cases where it was difficult to determine the sex of the individual, the Chancillería of Granada (a powerful judicial organism in Castile), in accordance with the law, permitted the ceremonial *electio* of sex, thus allowing Estebanía to choose: 'la mandaron que escogiese en el hábito que quería vivir e andar y eligió del de hombre' (she was commanded to choose the habit in which she wished to live and she chose that of the male). Such evidence shows how this process was not one of fiction, as some authors have tended to suggest, but one which was called upon in this kind of case.[15] Given the option to choose her sex, Estebanía chose the sex ('hábito') which allowed her the greatest social advantage.

Nothing is said, however, about the ceremony whereby Estebanía would promise to take on a particular sex (the 'juramento' (swearing by oath)), an act that had to be performed in the presence of the bishop and whereby the individual would commit to one sex and one sex alone for the rest of their days. As in the case of the Valencian draper mentioned in the previous chapter,[16] the judges in question seem to have overlooked this legal requirement. In this way, the institutional rite was frustrated or *misfortuned*, as Austin has referred to failed speech acts. The act by which authority confirmed in objective reality and collective perception, following Bourdieu, the official sex of the individual was invalidated in this case.[17] The fact that the procedure was not completed also suggests a less efficient set of bureaucratic steps than in the modern state scenario, and also an overlapping of administrative competencies from those of the royalty to those of the gentry and the ecclesiastical variety. Local, informal and traditional expressions of authority and law coexisted with official legal practice and norms.[18]

Estebanía, transformed into a man and now known as Esteban, found a woman with whom he entered into matrimony: 'se casó después con otra mujer e vivieron casados e velados en *facie ecclesiae*' (she married another woman afterwards and they lived as married and in *facie ecclesiae*). Her ability in the use of arms meant that she became a master of fencing, and she set up a school in Granada. As a way of 'ganar honra' (earning her honour), it is recounted that as Carlos V passed through Granada, Esteban was called upon to practise arms against some of the king's most worthy warriors ('hombres diestros y valientes' (brave and skilful men)). He beat them all: 'de los cuales batalló de todas las armas e los hirió e señaló él primeramente con la espada' (against whom he fought with all kinds of weapon and injured them and touched them first with the sword). When Esteban died relatively young, ten years after marrying, it is recounted as a thing 'notable de esta mujer hombre' (notable of this man-woman) that her mother and wife wept, the one for her daughter and the other for her husband: 'la una lloraba diciendo ay hija mía e la otra ay marido mío' (one cried saying oh, my daughter, and the other saying oh, my husband).

Estebanía managed to have her condition as a hermaphrodite and her transit to manhood accepted by means of a display of bravery, physical strength and her ability with weapons. These attributes were the province of a warrior ethos still highly valued at the time of the beginning of the Empire. They formed part of the myth of the *Reconquista* (Reconquest) and were valued as part of the ongoing colonization of the Indies.[19]

However, this ideal of military masculinity began a slow decline in the Spain of Carlos V in comparison to a new expression of courtly manliness.[20] This new development had appeared, for example, in the work of Baltasar de Castiglione, translated by Juan Boscán in 1534, precisely during Estebanía's early years.[21] This civilized, courtly and gentlemanly model of masculinity did not operate a complete break with previous modalities and continued to promote the skilful use of arms as one of its characteristic signs. The dexterity with which Estebanía practised fencing fitted perfectly with this new model.

Despite this, these skills did not fit comfortably with a new emphasis on the niceties of the spoken and written word, familiarity with legal procedures and the arts of intrigue and dissimulation. In this new aristocracy, military prowess began to take second stage to university training in the law or in the humanities. Rather than the externalization of aggressive impulses, the new manly model sought containment and restraint, a kind of ethos contrasted to the military antics that had exhausted the Castilian nobility in the early medieval period. During the military revolutions of the Renaissance, the role of this nobility was eclipsed by new techniques dominated by artillery, the development of the infantry (the famous Spanish *tercios*) and fortifications of one type or another.[22]

Given these changes, the conquest of the New World, where the knight could pursue the old bellicose model, represented a safe haven for outmoded expressions of masculinity, once removed from the influences of the court. It is possible that over the sixteenth century, this old model experienced a comeback in Spain as the old model of military and rural masculinity resisted renovation.

This, in turn, may have provided the basis for the delay in accepting the new type of subject proposed by Castiglione.[23] In any case, the Spanish experience in America served to mark the contrast between the metropolitan masculinity of the conquistador and the hybrid condition of the 'natives', both in the sense of blood and 'race' as mestizos and mulattos[24] and in respect of sexuality. In the latter case, the European imagination dreamed up nations of hermaphrodites[25] beyond the imperial shores and alluded to the frequency of sodomy[26] and later 'effeminate sodomites' among the indigenous of the New World.[27] This fascination for the military masculinities of imperial Spain effectively divests Estebanía of any charge of dishonourable activity, sinfulness, evil or bad portent. This master of sword-fighting is always described as an honourable villain. As such, she was declared to be 'claro de gesto' and was described as a 'persona bien nacida', something that suggested her condition of 'cristiano viejo'.

The covert allusion to 'blood purity' which distinguished the background of Estebanía from that of the *moriscos* – the fact that she had established a sword-fighting school in Granada, no less, should not be overlooked – and from that of the Jews means that her sexual peculiarities became related to questions of racial identity. But 'race', in the same way as 'sex', did not function as a bare biological category. We are still some way from the nineteenth-century concept of race and the arguments that led to the identification of 'poblaciones fenotípicas discretas e identificables' (discreet and identifiable phenotypical populations).[28] There was no simple biological identification of racial type; instead, one was subject to a network of interdependent characteristics such as one's guild, local community, age group and family lineage. In this way, the 'non-biological' attributes of Estebanía – in a word, her actions – earned her a page in the history books as 'heroic' and 'notable' rather than any truths derived from her anatomy.

Military prowess and heightened religiosity also fell into this category. Noble actions which were deemed prestigious were what engendered identity within a dense network of social relations.[29] This network of relations was not stable, inherent or created *ab initio*; it constituted an ongoing process of identification that depended on symbolic capital and on the meanings given by the representation that others gave those acts. The process of self-fashioning pursued by Estebanía was not one of a sovereign subject in voluntaristic terms.[30] What was operating was a form of subjectivity whose rules were not necessarily open to individual scrutiny at the time; it was vulnerable and depended upon the significances that others bestowed upon it.[31]

Estebanía, a kind of portent as described in the marvellous literature of the period, was undefined in respect of her physical sexual attributes but was identified by her conduct and the qualities for which she was socially admired; it was her military and religious qualities that bestowed upon her the honour of manhood in the age in which she lived.

Elena de Céspedes or the Misfortunes of Subalternity

The life story of Elena de Céspedes forms a complete contrast to what was an unequivocal expression of hermaphroditism in Estebanía. Historical interest in this figure has grown over the last few years, and Céspedes has become a kind of celebrity whose fame has gone beyond academic study to inhabit more popular genres such as the historical novel. Among the many works dedicated to her examination, we must distinguish between those of a medical nature[32] and those of a more historical nature, although the two are not mutually exclusive.[33] At the heart of such studies has been the attempt to determine whether Céspedes was an intersexual, a hypospadic male or a transsexual. Other studies focus on the picaresque aspects of her personality.[34]

The focus here, however, will be different. There is no attempt to pathologize the body of Céspedes, and the idiosyncrasy of this figure is not placed in a historico-cultural frame in an anecdotal manner. Our approach differs from analyses that place Céspedes's supposed extraordinary personal capacity of agency in the context of a heavily 'repressive' 'state' that tolerated no dissidence,[35] or those which argue in favour of a strong lesbian identity,[36] a 'proudly transgendered subject'[37] who defies a 'binary model of sex and gender'.[38] Instead, we will argue that the only way to understand Céspedes is to appreciate her within a context that was not dominated by this binary model, and instead to place her in a world that was wrought by tensions between the flexibility of magical natural representations (*mirabilia*) and rigid social and reproductive imperatives (*magicus*).[39]

Certain concessions along the way will be allowed for the subversive potentiality of Céspedes, but this is not understood as a voluntaristic process and within the framework of an unruly and transhistorical externalized personage such as 'the desire of the lesbian' or a 'transgender' identity in the face of normalization. Any subversive qualities shown by Céspedes are generated on the stage of a 'heterosexual matrix', it is to be admitted, but this possesses certain specific historical qualities whereby racial and sexual identities are fashioned not on the basis of independent biology but on the social forge of mutually constraining dependencies and modes of recognition. Such a network of discourses and practices turned Céspedes, as any other historical character, into a process rather than something static and not as a pre-modern identity already formed by tradition.[40] The unusual degree of creativity displayed by Céspedes in respect of her own life, furthermore, tends to question the standard account of the individual in pre-modern societies. In pre-modern or traditional societies, it is often thought, the individual is made up basically through the intervention of a number of collective influences such as parental descent, local community, class, age and the role of ecclesiastical and civil institutions. The modern subject, by contrast, would supposedly be conformed by means of his or her ability to construct their own identity.[41] Such a dichotomy needs to be questioned. Pre-modern identities have to be considered not as a combination of static and discreet components fixed by tradition and interiorized mechanically by individuals, but as a dense web of social dependencies in which individual biographies are inserted. The norms overseeing this process are what George H. Mead called the 'generalized other'.[42] The problem is that, as we have seen in the case of hermaphrodites and sex changes, these norms are not perfectly homogeneous, and neither is their action in any way automatic. What occurs is a performative reiteration of them by the body, and a space of indetermination is opened up.

Given this interpretive framework, the attempt by Céspedes to elude her condition as a woman – who was born a slave and was *morisca*, dark-skinned and poor – provides numerous difficult challenges for today's historian. Céspedes attempted something remarkable for her era: she tried to forge a biography con-cordant with the virtues and honours that at the time distinguished a person of 'quality' and which included, within this particular normative frame, the pos-sibility of 'improving' one's blood or sex. In some respects she failed to avoid abjection, and for this reason she was accused of sodomy, witchcraft, bigamy,[43] insulting the marital rite and transgression of the sartorial code. In her there is resistance, creativity and subversion, but these qualities are realized within what Butler calls a 'constitutive outside',[44] an interior process born on the basis of the norms of a particular sex/gender regime and heterosexual matrix.

Céspedes did not descend from 'Old Christians', despite what she alleged.[45] Neither were her origins *hidalgo*, as was the case of Catalina de Erauso. Céspedes was the daughter of a *morisco* slave woman, and as she was an emancipated slave herself her surname did not actually belong to her – she took the name of the wife of her owner. Only by means of an extraordinary effort to construct herself, modelling her body by means of a process of 'stylization',[46] and by elaborating the true story of her life, could Céspedes transgress the hierarchical borders between sexes and roles. She was a woman, a hermaphrodite, a man, a wife, a husband, a slave, a freed slave, a weaver, a draper, a shepherd, a domestic servant, a soldier and a surgeon.[47] Elena de Céspedes was able to do and undo her own personality many times over because the changing nature of commerce and the importance of particular skills, something amply illustrated by the comedies of the period in terms of identity 'fraud', permitted a degree of mobility that was greater than certain studies on the rigidity of social status prevalent in this society would sometimes suggest.[48]

The biography of Elena de Céspedes has been recounted many times, but it is worth concentrating on a number of details of her life. In 1586, in her forties, Céspedes lived as a man ('Eleno') in a small town near Toledo and was married to María del Caño, the daughter of a master artisan. She was some twenty years younger than Céspedes. In June 1587 Céspedes was detained by the *corregidor* of

Ocaña and was accused by a neighbour of committing sodomy with Caño. Shortly afterwards, she was summoned by the Inquisition and was accused of insulting the sacrament of marriage and of being involved in acts of witchcraft. The jurisdiction of the case fell finally to the inquisitorial office of Toledo, which, among other things, freed Céspedes from the charge of sodomy and possible death.[49]

The main argument that Céspedes used to escape any charges was that she was a hermaphrodite. This quality, she argued, undermined the accusation of disrespect towards matrimony since, although she had indeed been married to a man as a woman, it was only on the appearance of male genitalia after giving birth that she felt inclined towards the female sex and then married María del Caño. This argument also quashed the charge of witchcraft, which had arisen because Céspedes was accused of having deceived those who had examined her to prove her status as a male on several occasions, and in particular just before her second marriage. Among those supposedly deceived was none less than Francisco Díaz, surgeon to Felipe II.[50] Díaz had examined Céspedes before her wedding on the request of the Curate of Madrid, and he had declared her to be a male. When Díaz was called before the *Santo Oficio* and inspected Céspedes once more, he stated that she was a woman who showed no trace of manhood. Finally, the tribunal of Toledo found that they were confronted with a case of fraud. Céspedes had always been a woman, they believed, and they pronounced a sentence similar to those given in cases of bigamy. This entailed a public *auto de fe*, two hundred lashes and ten years' reclusion and service as a doctor in a poor hospital.

Céspedes never denied that she had altered her body in order to hide her female genitalia and to emphasize her male nature. But she sustained right to the end that she possessed both qualities, although her male member, according to her, had begun to diminish and decompose some time ago to the point of disappearing during her stay in the inquisitorial prisons.[51] Elena de Céspedes was able to convince so many people because she gave herself a particular identity, which was constructed by means of an account that combined her own readings as a self-trained doctor (her library was extensive)[52] and her experience, which was flushed through with practical know-how and oral traditions, something typical of *morisco* medicine.[53] Céspedes knew how to combine these diverse elements in order to provide a convincing case before whichever authority (ecclesiastical, medical, magistrates, neighbours, etc.) she appeared.

This degree of creativity was only possible in a network of discourses and practices such as those present in the *Ancien Régime* of sexual identity where the three kinds of experience described above occurred. Firstly, especially in the discourse of Céspedes herself, value was given to knowledge of medicine and natural history in order to present her case as something strange, a 'preternatural' event, but never beyond the bounds of Nature. In this way, she managed to dodge the charge of sodomy, despite its omnipresent possibility during her trial.

Secondly, Céspedes emphasized her occupation as a soldier and her participation in the repression of *morisco* uprisings. But in her case, unlike that of Estebanía, recourse to her ability to wield arms did not convince. Unlike Estebanía or Catalina de Erauso, it was Céspedes's sexual practices that cast doubt on her military prowess. Céspedes, like Erauso,[54] led an itinerant life; but in contrast to the latter, she found herself in a kind of 'no man's land' whereby she became almost a vagabond, the misfit *par excellence* of the old regime.[55] Despite her efforts, this situation made her into the counter-image of the virile *conquistador* inhabited by Erauso. In addition, her *morisco* background reinforced the negative associations entailed by this quality. It was as though Céspedes was a kind of bandit (a *monfí*) that roamed the hills of the sierra around Granada.[56] In a climate that accused these bandits with uprisings in the villages and small towns of the Alpujarras, with Berber incursions against the coast and with the dangers represented by a possible Turkish invasion, Céspedes became irremediably associated with a biopolitical threat to the order of the Spanish Empire.[57]
· Her connections with Moorish medicine further associated her with those practices, especially witchcraft and pacts with the devil, which the Inquisition combatted.[58] Her skin colour reinforced her 'non-Europeanness' and suggested her descent from those exotic lands populated by fantastic beasts and hermaphrodites located by the sea-faring naturalists of the period in India and Ethiopia.[59] Her shift from wife to husband, her multiple partners, her seduction of married women and deflowering of damsels, her disrespect towards the marriage sacrament and, finally, her actions tainted with witchcraft moved her perilously close to evil terrain and sin *contra naturam*. In fact, the inquisitorial tribunal considered it an act of clemency to have freed her from the charge of sodomy as brought by the Royal Tribunal.[60]

The charge of sodomy merits especial attention. First of all, Céspedes's *morisco* background associated her with the sin of sodomy. A literary and historical tradition culminating in the reasons provided for the expulsion of the Moors in 1609, among other acts, accused them of readily practising sodomy.[61] In the case of Céspedes, this would be a particular form of sodomy. The accused defended her status as a hermaphrodite throughout her trial, and this condition would have allowed her to engage in sexual intercourse with her short-lived marriage to the stonemason from Jaén and with many women.

From the juridical point of view, as was pointed out in the previous chapter, hermaphrodites were not immune from being accused of sodomy. Those who possessed a predominant sex, whether male or female, were considered as men or as women in respect of their sexual life. Those that were 'perfect', possessing the natures of both sexes, should choose one of these and swear before a bishop that they would remain true to this identity in order to ensure their heterosexuality for the rest of their lives. Céspedes argued implicitly that her particular type of hermaphroditism belonged to the 'hidden' category, a concept that, as we have seen, was pertinent to medical discussions on sex change. Hidden hermaphrodites revealed an ambivalence that had arisen by accident as part of the process of 'improvement' or change from woman to man; in Céspedes's case, the virile member would have subsequently disappeared once she was in prison.

The strategy followed by Céspedes, therefore, consisted in arguing that her form of hermaphroditism had remained hidden or had reverted back to this status and that therefore she could not be judged in accordance with the standard procedures employed by the legal authorities. This allowed her, just as had happened in the case of the Valencian dressmaker discussed by Matheu i Sanz,[62] to avoid the accusation of sodomy by relying on a possible bodily defect. Even though she did not manage to convince the tribunal of her hermaphroditism, she was able to convince her tormentors that it should be the Holy Office alone that would deal with her case, and that the charge should be bigamy and insults to the sacrament of marriage rather than the more serious charge of sodomy.

As the trial proceeded, it became apparent that Céspedes, as a woman, had penetrated other women and that she had done so by using home-made dildos or 'baldreses',[63] probably covered in animal skins. Drawing on her own medical knowledge and readings in natural history, Céspedes attempted to refute this charge and suggested that her virile member was the result of her peculiar physical qualities: she was a woman who possessed a large clitoris, or what has been termed a macro- or hypertrophied clitoris.[64] Spanish doctors, for example, García Carrero in his *Disputationes medicae super fen libri primi Avicennae* (1611), had argued for the existence of this intermediate figure in the sex/gender hierarchy. Manly women viragines were capable, it was thought, not only of penetrating other women but also of possessing sufficient heat to enable them to ejaculate sperm that could impregnate their partner. Although this latter opinion was not accepted by most medical figures, the question of the macroclitoris certainly was approved by many.[65]

In the penal typology of the period, the possibility of female sodomy was accepted. Trials for this crime were, nevertheless, exceptional in comparison with those for male sodomy. In the prevailing juridical framework, which subscribed to a broadly phallocentric understanding of sex, there existed an 'imperfect' form of sodomy among women involving caresses and masturbation. The Latin term *mulier fricatrix seu subigatrix* was used in order to designate women who performed these acts.[66] In practice, however, these acts were generally not sanctioned, if we are to believe various literary documents, including the comedies *Añasco el de Talavera* (c. 1637) by Cubillo de Aragón and *El sueño de la viuda* by Melchor de la Serna (c. 1606),[67] and the account by María de Zayas, *Amar sólo por vencer* (1647), which shows how erotic attraction between women was deemed ridiculous or served as a pretext for a reflexion on Platonic love rather than any sinful characteristics.[68]

Any penetration of another woman with a 'material instrument', however, was a different matter. This act was equated to male 'perfect sodomy' and was in principle punishable by death. However, in those cases in the Spanish ambit that have been studied to date, for example that concerning Inés Santa Cruz and her companion, Catalina Ledesma, no death sentence was passed.[69] As it was believed that the semen that was spent in these cases did not have any consequences for procreation, divine creation was not undermined and the disorder caused was far less important than was supposed by similar male activity. It is possible that this is the reason behind the fact that the inquisitors of Toledo did not remit the Céspedes case to the royal Castilian authorities in order to try her for sodomy.

Finally, the case of Céspedes also has recourse to the experience of miracles, of the hermaphrodite as an individual equipped with redemptory and fantastic powers.[70] It is only in this knowledge that we can explain the commotion that the presence of Céspedes caused in the Hospital del Rey in Toledo, to where she was sentenced.[71] There were stampedes in favour of being treated by the surgeon from Alhama, causing disturbances and a series of problems that could only be resolved by transferring Céspedes to another hospital. She went first to another hospital in Toledo and later to the village of Puente del Arzobispo. Here, the traces of this exceptional figure fade away, although the memory of her person lasted for nearly a century.[72]

Catalina de Erauso and the Crisis of Imperial Masculinity

Catalina de Erauso, born in 1592, when the trial against Elena de Céspedes had already finished, led an even more itinerant lifestyle than her countrywoman.[73] Her unusual life story has inspired a range of literary and artistic creations, and she has been the subject of many studies. Most recently, studies have emphasized her apparent ability to challenge the sex/gender schema of the period which has been presented in dichotomous and essentialist terms; Erauso would in fact create her identity from a set of normative moulds which were subject to the frameworks provided by the dominant male and female norms of the period.[74]

In addition, numerous studies have focused on the interconnections between Catalina de Erauso and the qualities ascribed to the perfect male and to heterosexuality. Erauso's sexual preference was for women, but she remained a virgin; and this research has outlined the context of the Golden Age and argued that sexed identity was not a reality fixed by biology in an essentialist manner, but that it depended on the recognition that she received within a dense network of social relations. The lower noble condition of Erauso as an *hidalgo*, the family's social capital, her heroic military acts, her Old Christian religiosity – all allowed Erauso to externalize her 'inner moustache' and be recognized as a male by the pope and the king and to rise in the social hierarchy, which was also the hierarchy in terms of sex.[75] Once again, it is necessary to calibrate carefully the level of agency of Erauso, not by means of some unruly and ahistorical 'non-phallic desire' but by recourse to an understanding of the norms operating within the heterosexual matrix of the time, which guides not only ideal conduct but also that which is abject.[76]

On escaping from the Dominican convent in San Sebastián at the age of fifteen, a place her parents had sent her in order to become a nun, she began a long personal journey through the territories of South America, passing through Colombia, Panama, Peru, Chile and New Spain. She returned to Europe on a number of occasions and made longer stays in cities in Spain and Italy. She exercised as many trades or more than Céspedes. It was not only arms, the activity that made her famous as the 'nun ensign', that Catalina would take up; she also worked as a page, cabin boy, servant, butler and tax manager of diverse activities in mining, livestock farming and merchandise. She altered her identity on several occasions. She changed name from Catalina de Erauso to Francisco de Loyola. When she enlisted in the militias, she became Alonso Ramírez Díaz and gained the rank of lieutenant. When, in 1620, she was forced to confess her true identity to the bishop of Guamanga, she revived her first name and condition as a nun, but changed again when she received permission from the king and the pope to dress as a soldier and to receive a military pension as Lieutenant Catalina de Erauso. As such, she finally went to New Spain under the name Antonio de Erauso.

On first sight, the seriousness of Erauso's misdemeanours exceeds that of Céspedes. In several duels and fights she killed eleven persons of varying stations, including a slave, a bailiff and a beggar. She was condemned to death on more than one occasion and on others could only save herself by going into hiding. In addition, despite her pretensions to virginity, she was involved in various amorous affairs with young women and even with a married lady. With this reputation as a gambler, murderer and adulteress, how is it be explained that Catalina de Erauso was not only capable of becoming accepted by the authorities – receiving the blessing of king and pope – but also managed to become a popular legend and, at the same time, an exemplar of the perfect Spanish *caballero* and the perfect Spanish lady?[77]

One crucial element is clearly the military *habitus* occupied by Erauso. She came from a Guipúzcoan family and shared the *hidalguía* and corresponding privileges accorded to all Basques. She was famous in military milieus, and her reputation was recognized in the militia, the navy and among the clergy.[78] The decision to become a man and to enrol in the campaigns of the Spanish army overseas was not taken in splendid isolation; it was taken in the context of a solid military tradition prevalent among Basque nobles. The fact that Erauso, without revealing her true identity, was involved in such activities and was accompanied by family members and associates in these activities was not arbitrary. Such links afforded her cover and also a support network that enabled her to achieve certain things, such as a pension from the king.[79] On this basis we can understand how Erauso incorporated a military *habitus*, conceived as deriving from the tradition of the Reconquest.[80] Even if she lost her position of prominence with respect to the court in Madrid,[81] she continued to play a fundamental role among the *hidalgos* who took part in the wars overseas.[82]

The admiration for the 'natural pendenciero' and bravery of Erauso,[83] inherited in the blood from her ancestors but common to the imperial age of Spain as part of a notion of the need to defend honour and *hombría*, corresponds to a form of social organization in which the state did not hold the monopoly on violence. Duels and blood contests,[84] daily events in peninsular towns, were even more common in colonized lands. These territories were characterized by a 'frontier mentality', where authority encountered all kinds of obstacles.[85] In this kind of 'civilizing' context,[86] survival depended on physical strength and the ability to use arms. It was here that the 'Monja-Alférez' excelled; her murders were justified on the basis of the restoration of honour, and it is for this reason that she managed to fend off any condemnation for her acts triumphantly.

The exaltation of the manly qualities of Erauso did not take on the same lustre as that seen in the case of Estebanía de Valdaracete. In her case from the mid-sixteenth century, the slow process of 'acortesanamiento' of the nobility was still unfolding. In Erauso's time, at the beginning of the seventeenth century, this process was already well under way. The figure of Don Quixote shows how time-bound the subjectivity of the military knight was as it was steadily supplanted by an urban, courtly, bureaucratic aristocracy.[87] For this reason, the scope of manliness in Spain is not contrasted merely with the alterity of the New World or with racial or sexual hybridity or even sodomy. The contrast is not with something external but rather something internal. The debate over masculinity has shifted to what is taking place within the realm in the context of the new thought on the governability and the economics of the nation in Europe. In light of the decline of the empire and the imperial subject,[88] the space that is held up for scrutiny is the court itself. The focus is centred on the supposed effeminacy of the nobility, as the model shifts from a warlike caste to one based on courtly values. In turn, the imperial crisis was presented as a crisis of masculinity. Catalina de Erauso operated in effect like an instrument that allowed for the criticism of the nobility and their lack of bravery at times of imperial need.

This concern merits a few lines of analysis. Firstly, a whole group of *arbitristas*, from González de Cellorigo (1600), Quevedo (1609), Pedro de Valencia (1618), Sancho de Moncada (1619), Pellicer de Tovar (1621) and Fernández Navarrete (1626),[89] through to moralists, such as Fray José de Jesús María (1600), Suárez de Figueroa (1621), Francisco de León (1635), Íñigo de Camargo (1689) and Fray Antonio de Ezcaray (1691),[90] and commentators on fashion and dress, including Fray Tomás de Trujillo (1563), Fray Tomás Ramón (1635), Alonso

Carranza (1636), Marqués de Careaga (1637), Jiménez Patón (1638) and León Pinelo (1639),[91] critiqued courtly (re)finery, seen as the cause of the effeminacy that was destroying the Spanish nobility.

It was believed by these individuals that men were spending too much time on their physical appearance than women were themselves. Men's preferences were, moreover, 'womanly', as evidenced by clean shaving, curls, the use of perfumes and the use of showy or colourful vestments.[92] There was something 'trans-natural', in the words of Rappaport, about this form of dress and demeanour – an aspect that was often pointed out in cases of sex change.[93] Such exemplars of the nobility, on dressing themselves up 'as women', literally became them, just as the donning of male attire changed Erauso into a male officer. The first case scenario resulted in an ontological decline or decay into an inferior being. The case of Erauso produced an improvement in rank. Such a loss in 'quality' could affect both the sexual hierarchy (the change from man to woman) and the hierarchy of blood (from Old Christian to New), as in both cases there was a decline in one's rank. This also explains why the female attribute of menstruation became associated with Jewishness. In some cases, certain doctors such as Andrés de Laguna, Juan de Quiñones and Isaac Cardoso argued, it was possible to identify false converts in men by means of this quality.[94]

This concern about 'effeminacy' did not inhere necessarily in the sexual realm.[95] Not all effeminacy spelled sodomitical desire or practice, and not all sodomitical activity signified womanly traits.[96] The 'effeminate sodomite' that proliferated in colonial Mexico in the 1650s,[97] associated with the passive role in intercourse, was only one sexual persona among many in the sexual order of the period. Effeminacy was linked to weakness, to a degree of softening that resulted from men having abandoned military positions and from the growth of courtly masculinity. In a society in which identity was the effect of actions and not the reverse, the cultivation of the spoken and written word, the seeking of intrigue, and dance and literature all tended to point towards the creation of an 'effeminized' form of *hidalgo*. Extramarital sexual activity, sodomy included, was also tainted with weakness, lack of restraint, decadence and excessive luxury.[98] A similar concern is reflected in the literature written by the *arbitristas*. Such works highlighted economic decline, the frittering away of the precious metals discovered in the Indies and the purchasing of exquisite foreign goods. This desire for what was deemed superfluous was compared with extramarital sex: men suffered from 'love sickness',[99] abandoned their wives, neglected their marital duties, sought lovers outside of marriage, took mistresses or frequented brothels.[100] Luxuriousness, female seductiveness and extramarital passions debilitated men and converted them into passive subjects that consumed.[101] In order to shore up this degenerative tendency, the old values of the austere rural nobility were invoked, although they were incorporated into a new model of self-contained aristocratic virtue. In philosophical terms, this shift was grounded in the thought of Seneca and in a recuperation of the thought of the Stoics.[102]

Those in power, Fernández Navarrete argued in his *Conservación de Monarquías y Discursos Políticos* (1626), should avoid the easy life and sensualism in order to impede the masses, in their rush to imitate such lifestyles, from abandoning decent marital relations, thus 'quedándose con un celibato poco casto' (ending up with a form celibacy that was rather unchaste).[103] In the final analysis, the fall in population levels was in part due to the effeminization process that had trickled down from the upper classes to the mass of the people. In turn, depopulation was at the heart of the decline of the empire and Spanish power.

It was these concerns, among others, that led the Grand Duke of Olivares and others to establish, shortly after Felipe IV took the throne, the Junta Grande de Reformación in 1622. The Junta approved a series of measures designed to encourage marriage and to address the moral decline in the nobility, a situation that had reached its zenith during the period inhabited by powerful figures such as the Duque de Lerma under Felipe III.[104] Among the measures was the *Pragmática* of 1623, known as the order of the 'recently weds', which sought to encourage marriage and prevent arrangements that sought to merely increase the wealth of the partners. It also provided tax relief for families with more than six children and penalized bachelors over the age of twenty-five. The same legislation included provisions for the closure of brothels.[105] Olivares, in his *Memorial* on education published in 1632, criticized the lack of military vocation in the nobility and youth of the day.[106] In this sense, Catalina de Erauso served as an example of the conservation of the old values even though they were encapsulated by a woman who had fought for the conquest of the Indies. As such, Erauso was popularized in Spanish society at the time, not least given her presence in comedies such as *La Monja Alférez* by Pérez de Montalbán (1626).[107]

The other sphere where this concern over Spanish masculinity was expressed was the literary field. As numerous recent accounts have shown,[108] a growing critique of effeminacy found its expression in the theatre of the day, especially in short theatrical works which ridiculed 'soft' masculinity in the form of 'lindos' and 'mariones', exemplars of womanly men. In addition to this kind of piece on the mannerisms of noble men, a second type of theatrical work represented 'hidden' men, that is, men who dressed as women or who were sufficiently ambiguous in order to hide their male identity. A third set of examples depicts amorous relations between men; here effeminacy was not present and transvestism was not always the case.

Those works critical of 'soft' masculinities included comedies such as *El labrador venturoso* (c. 1635) and *La bella Aurora* (c. 1635), both by Lope de Vega, pieces that attack the figure of the 'lindo' and associate this demeanour with sodomitical practices. *El Narciso en su opinión* (1625) by Guillén de Castro and its later version by Agustín Moreto, *El lindo Don Diego* (1662), followed suit. The protagonist of this work imitated the extravagant and effeminate mode of dress of the Madrid court.[109] Similar to these pieces are those that dealt with 'mariones': *Los Mariones* (c. 1645) and *El Marión* (c. 1645) by Quiñones de Benavente, *Los Maricones galanteados* (c. 1674) by Gil López de Armesto and *El Marión* (c. 1628) by Quevedo all fall into this category.[110] These all parody their subject by depicting women who challenge one another to duels and young cloistered men who exhibit female behaviour but who still dress 'as men'.[111]

The second set of works, that of men dressed as women or who played the role of 'hidden' men, allows for a certain degree of confusion in seduction scenes. Short works or 'entremeses' – such as *La Gran Sultana Doña Catalina de Oviedo* (1615) by Cervantes, *La dama fingida* (anonymous; last third of the seventeenth century), *Pistraco* (c. 1640) by Quiñones de Benavente, and the dance by the same author, *La casa al revés y los vocablos* (c. 1645), which fantasizes on a carnivalesque 'world turned upside down' – are examples of this device.[112] Pieces representing 'Juan Rana', played by the well-known actor Cosme Pérez, complete the selection.[113]

The third type does not simply include those male characters that were dressed as women. They suggest a whiff of desire between men. This can be seen in two pieces by Quiñones de Benavente: *Turrada* (c. 1645) and *La Hechicera* (c. 1645). This trope is also present in *Los Putos* (c. 1651) by Jerónimo de Cáncer, as well as in comedies such as *La Boda entre dos maridos* (c. 1600) by Lope de Vega and *Los Empeños de una casa* (c. 1683) by Sor Juana Inés de la Cruz.[114]

In general, this theatre corpus can be included in the carnivalesque tradition whereby hierarchical social roles are inverted with the intention of questioning prevailing social customs. What is not clear is whether this type of play merely wishes to highlight the degeneration of the masculine model or whether it comprises the hope and expectation that masculinity will be regenerated in the future.[115] What is certain is that we should not think, as is too often the case in certain branches of gender studies, that the appearance of men dressed as women in the theatre of the Golden Age automatically supposes a subversive gesture or a way of legitimizing the existing sex/gender divide. The significance of these performative acts always depends on the context in which they take place. In this way, 'transvestism' may well be a simple narrative device in the play's story.[116]

In the daily life of the period, this carnivalesque inversion of sexual attributes may not have been absent. This was the case of the burlesque marriage celebrations that took place in the court in 1623 and in 1638, when the participants appeared dressed in the clothes of the other sex, or in certain popular festivals.[117] In general, however, this alteration of sartorial styles was viewed with some suspicion. For this reason, authors such as León Pinelo in his *Velos antiguos y modernos* (1639) asked for the Castilian laws against the use of masks outside of festivals[118] to be extended to sanction men who dressed as women.[119]

Norms on theatrical representation also reflect these concerns. Even though women dressed as men on the Spanish stage were not necessarily viewed positively and may have been subject to censorship, men dressed as women were a different category altogether.[120] The usual plot entails that women take on men's clothes in order to restore their honour, an act that converts these female actors into heroines. By contrast, the comparatively fewer plays that depict men dressed as women generally tend to ridicule male effeminacy. These rules governing cross-dressing on the stage would seem to mirror those employed by doctors and natural historians: the fact that a woman dressed as a man could be interpreted exceptionally, as in the case of Catalina de Erauso, as an attempt to improve status and rank. The reverse was considered dishonourable.[121]

Novels also reflected the same kinds of concerns about effeminacy in men in the context of the rise of courtly manners.[122] *El Andrógino* (1622) by Francisco Lugo y Dávila narrates the love affair between two young people, Ricardo and Laura, which is problematic because of the social status of each party.[123] Laura, born into a lower-class family, was married off to the older and wealthy Solier. Solier confined his new wife to his Valencian mansion, which became an impregnable fortress retaining Laura. In order to gain access to the house and to recover his lover, Ricardo dresses as a woman and passes himself off as Bernardina, a fallen woman whose plight moves the old Solier. The latter falls in love with Bernardina and decides to take her in so as to provide some female company for Laura. Ricardo was thus able to continue to see his loved one until Solier finds them in the same bed and discovers that Bernardina possessed a male sex. Ricardo pleads that whether as a result of a miracle or through Nature, he has changed into a man. Astonished by such a development, Solier rushes to interview Salt, a professor of medicine at the university. Salt, on consulting with his colleagues, launches forth on a detailed and erudite lecture on the possibilities of sex change from woman to man and on the impossibility of the reverse transformation.

The critique of womanly masculinity is even more explicit in the works of María de Zayas y Sotomayor, such as *El Decamerón Español* (1637) and *Desengaños amorosos* (1647). In the first of these, the beautiful Lisis, a novelist and alter ego of María de Zayas herself, attacks men for their diminished manhood and virility. Men have forgotten, she believes, their obligations in respect of lineage, the nation and their own status as men and have given themselves up to passion, fashion and superfluous pleasures.[124]

One of the symptoms of this 'devirilization' would coincide, as *Desengaños amorosos* suggests, with misogyny, the cruel treatment of women and their abandonment by men. One of the stories in this series, 'Mal presagio casar lejos', relates the abandonment of Doña Blanca by her husband, a noble who had succumbed to sodomitical passion with his pageboy. Neither of these individuals was dressed as a woman or was particularly effeminate, but both were caught by the wife in a sodomitical embrace in the conjugal bed.[125]

María de Zayas critiques in this way the attitude of the *hidalgo* of her time and his lack of observance of the modalities of the period: his impetuousness, his lack of interest in clothes and fashion, and his strong will. The violence suffered by women, their lack of education and their second-class social status as 'mujer eunuco' (woman eunuch)[126] is mirrored by the degeneration of the man, who does not fulfil his obligatory role of dominance in social and sexual relations. The contradictions of the 'proto-feminism' of María de Zayas have been discussed at length whereby the subordination of women was critiqued but the values of male-dominated society were left intact. In fact, the author of *Desengaños amorosos* was only able to formulate strategies that emerged as part of the androcentric order itself and was limited to appealing for a kind of masculinity that fitted with the obligations already established under the rubric of the nobility. Both masculinity and nobility, therefore, were to be in effect consolidated by such a move.[127]

Men could only exert their 'natural' dominance over women if they could effectively display their power and superiority in the social field. This superiority clearly was called into question when masculinity was placed in jeopardy. For this reason, the work of María de Zayas exalted active and voluntaristic women who controlled the lives of their husbands,[128] or who were prepared to test their resolve and courage on the battlefield, as in the case of the female captain Estela, the protagonist of 'El juez de su causa', a story included in *El Decamerón Español*.[129] It is in this light that we can understand Catalina de Erauso, a woman who fought on the terrain of a devastated masculinity and sought to reinstate virility and gain status through her actions.

But Erauso was not devoted wholly to war-like acts of male bravery. She was also aware that she had to incorporate one of the qualities that were essential to the perfect woman in Counter-Reformation Spain: that of virginity. After killing a bailiff, a black slave and an official who attempted to seize her, Catalina sought protection under the authority of the bishop of Guamanga. After confessing her status as a woman, she was examined by two midwives, a doctor and a surgeon. All of them confirmed her virginity:

> Como a las quatro de la tarde, el mismo Señor Obispo abrió la puerta y entró. Entraron detrás dos mujeres, que eran comadres y un Médico y dos Cirujanos. Mandóle a todos con censuras reservadas, que hicieran su oficio legalmente y salióse fuera y cerró. Yo me manifesté. Ellos me miraron y se satisficieron de que verdaderamente estaba virgen ... Su Illma. se enterneció y allí delante de ellos se llegó a mí y me abrazó y me dixo: Hija, ahora creeré todo cuanto me dixéreis. Yo, con humildad y reverencia me arrodillé y le besé la mano.

> (At about four o'clock in the afternoon, the Bishop opened the door and entered. Behind him, two women entered who were midwives and a Doctor and two Surgeons. He commanded all with discreet reserve that they perform their legal offices and he left and closed the door. I uncovered myself. They examined me and were

satisfied that I was a virgin ... His Excellency was touched and before them all he approached me and embraced me and said: My daughter, now I will believe everything you say. With humility and reverence, I kneeled and kissed his hand.)[130]

This quality is referred to emphatically in Erauso's account. Virginity was proof of virtue,[131] a state that made up for her transgression of sexual boundaries. Although this transgression had been achieved by a long and painful process of 'styling' her body,[132] it mitigated the blame of her amorous activities with other women and exonerated her from the accusation of sodomy. Her reputation as a virgin was, it would seem, one of the main reasons for her popularity at the time.[133] This woman, steeped in virility and military bravery, was at the same time a true 'shrinking violet'.[134] The virgin warrior, who had become famous after her voyage to Rome where she kissed the feet of Pope Urbano VII, also showed clear signs of impeccable religious devotion.[135]

Combining the war-like ardour of a soldier and the candour of a young lady, Erauso was able to construct her own myth in order to become, as Elisabeth Perry has argued, a metaphor of the Spanish realm.[136] As a result of her status as a true oxymoron, the Monja-Alférez could bridge the category of God-given marvels and that of an almost miraculous virginity conserved in the context of violent masculinity. In this way, any connotation of evil linked to committing the nefarious sin, something that was palpable in the case of Elena de Céspedes, was instantly dismissed. Catalina de Erauso was able to change not into a male but into a legendary symbol of Spanish patriotism, a kind of tough ancestral version of Agustina de Aragón.[137]

Juan Díaz Donoso and the Sex of the Clergy

Our next case takes us to a very different setting. The moral universe is no longer that of the dimished masculinity of the Spanish court but that of the Christianizing campaigns of the Counter-Reformation in the wake of the Tridentine decrees. The chief internal objective of such reforms was the clergy itself. Together with the concern over the lack of intellectual preparation of the clergy and what was perceived to be their indulgent lifestyle, the question of sexual morality was high up the list of matters needing reform. The elimination of concubinage, the tracking down of wayward priests, a growing condemnation of voluntary pollution, the control of 'particular friendships' and the frequency of sodomy in religious houses all moved centre stage.[138] Although it took some time for this strategy to take effect, it represented greater vigilance over and disciplining of the life of the clergy.[139]

The charges against the priest Juan Díaz Donoso from Zafra are relevant in this context.[140] The Inquisition file that has been preserved pertaining to his case is incomplete (the resolution from the Holy Office of Llerena is absent) and suggests a hearing that was interrupted. It would appear that the case was passed to the jurisdiction of the diocese of Badajoz. What has remained allows us to gain an impression of a panorama that to date had been little explored: the attitude of the ecclesiastical and medical authorities towards hermaphroditism and towards the sexual question in the life of a member of the clergy.

As Soyer has pointed out, the investigation by the Inquisition was sparked off by the declaration of the shoemaker, Domingo Rodríguez, against Juan Díaz Donoso on 24 February 1633, accusing him of having maintained sexual relations with another man. This other man, 'Juan el Portugués', a shoemaker's apprentice, had been seduced by the priest some days earlier. Juan confirmed to Rodríguez that he was not the first to have been seduced in this manner; others included a young dressmaker from Badajoz. What is perhaps most interesting about the case is that Juan declared that in their carnal relations the priest had 'acted like a woman'. For this reason, Juan insisted that the devil had deceived him into performing sexual acts with the priest. Sceptical of this possibility, Don Alonso de Jeremías, in charge of the inquisitorial office of Llerena, indicated to his superiors that Donoso was conceived as a 'mujer ermafrodita' (hermaphrodite woman).

What, precisely, was the nature of the crime committed? If it was a case of sodomy, jurisdiction did not correspond to the Inquisition but instead to the secular court. On the other hand, given the fact that the protagonist in question was a priest, it was the episcopal authorities that should hold sway. If Donoso was in reality a woman, his/her taking up of the cloth could encourage his flock to believe that women were capable of being ordained and fulfilling the role of priest. Donoso had also even proposed marriage to the son of a tax collector, a possibility that was prohibited for the clergy. Given that these contraventions of faith had apparently taken place, the Inquisition could be legitimately involved in the investigation of the case.

Matters became even more complicated once the accounts of witnesses were heard. These included a former mayor, a lawyer, Juan el Portugués, the tax collector and his son, a notary, a villager and a female servant who were additional witnesses to those who had declared before the Diocesan Tribunal of Badajoz (these, in turn, included four people from the locality, a dressmaker and a pastor), which had taken up the case in February 1635.

It became evident that the priest had shown many of his potential partners a document attesting to a papal dispensation in order to convince them to sleep with him. One witness, who was unable to read the original Latin, was told by Donoso that the text allowed him to choose the 'estado que quisiese, de hombre o de mujer' (state that he wished, as a man or as a woman).[141] Pedro Alonso, the son of the tax collector, was shown the dispensation and was told that Donoso had been bestowed the privilege of marrying or electing the sex he desired.

It was complexities such as these that made the Holy Office in Llerena seek outside assistance and refer to the Supreme Tribunal of the Inquisition in Madrid. In turn, on 20 November 1634 the Madrid office asked for advice from the prior and brothers of the convent of Nuestra Señora de Atocha, the most important convent that the Dominicans possessed. The guidance provided by the brothers has not been recorded or is lost. However, it was noted that rumours about Donoso had been current for nearly a decade and that some believed he was a woman and a man at the same time, while others believed that he was a woman only. Others still remarked upon his 'effeminate and womanly' traits or signalled the devil as the architect of the priest's behaviour. Donoso himself contributed to this confusion by arguing on occasion that he was a woman who appeared to be a man. He had told Juan el Portugués, for example, that he knew of certain potions that would make a man's beard grow. In this sense, Donoso applied, like Céspedes and Erauso, a certain 'stylization' of the body. On other occasions he also mentioned that he possessed both sexes (in a conversation with Alonso Delgado), was exclusively male (when intimidated by threats from pastor Manuel Carvalho) or had changed from a man into a woman (in a conversation with Pedro Alonso when he stated that his small penis 'se marchitó y fue sustituido por una vagina' (shrivelled up and was substituted by a vagina)),[142] thus using an argument similar to that employed by Céspedes in her defence.[143]

Rumours abounded on the sexual predilections of the priest for young men. Young men were regularly invited to stay at the priest's house, where he entertained them with music and song. On one occasion he was beaten up by a member of the local community and accused of being a 'whore' and of jealously coveting Alonso as if he were the priest's husband. Having been rejected on numerous occasions by Pedro Alonso, Donoso lay in wait for him, armed with a sword. Despite this, Donoso cannot be compared in this sense Erauso; no violence was committed.

The agents of the Inquisition believed right from the start that before them they had a case of female hermaphroditism, and that because of this Donoso was not capable of performing the role of a priest (he was not 'capaz del carácter sacerdotal'). In this sense, it was a question of ecclesiastical competency. The inquisitor of Llerena, Juan de Vallejo y Acuña, recommended in February 1635 that Donoso be arrested and taken to the prison of the Holy Office. He suggested to the Supreme Tribunal that his condition as a hermaphrodite be examined carefully by designated medical men. These experts should declare as to the organs that the priest possessed and as to whether he had employed them for sexual purposes. It was also recommended that two midwives were present for the examination.

The last document in the inquisitorial file on this case is a letter sent to the Supreme Tribunal by Vallejo y Acuña on 12 April 1635. In this missive, Donoso is referred to as 'un hermafrodita residente en Zafra', and he remarks that the Holy Office in Llerena has now made its decision on the case. This decision, however, has disappeared, and any remaining traces of the process have been lost.[144]

What remains of the documentation on Donoso shows clearly the interest shown by the Inquisition and the ecclesiastical authorities in general in the lives of its subjects and, in particular, in the sexual lives of the priests of the Church. This will to truth was particularly intense at the end of the sixteenth century in Extremadura, and especially in Coria (Cáceres). Here, the episcopal authorities obliged the inhabitants and the priests of the locality to engage in a non-sacramental confession detailing their vices and excesses. At the top of the list were sins related to luxuriousness.[145]

Perhaps the most interesting aspect of the Donoso case is the supposed papal dispensation. As discussed in the previous chapter in respect of the juridical statute of the hermaphrodite, it was not possible for the individual to choose freely his or her sex or to get married. As we have argued, *electio* according to sex could only take place in those cases where there was no predominant sex in the hermaphrodite. The individual concerned had to swear before a bishop that this sex or *natura* would be kept to throughout his/her life. The Counter-Reformation had ensured that celibacy, chastity and the impossibility of marriage would be binding for the clergy.

What did this dispensation, if that was what it was, actually mean? In order to understand the situation of the hermaphrodite and the possibility of taking religious vows and becoming a priest, it is necessary to discuss the canonical law of the period. Some commentators have wished to interpret certain cases of hermaphroditism and sex change that took place in convents as the main motivation behind attempts by theologians such as Martín de Azpilcueta to draw up a set of rules to govern the status of hermaphrodites in the ecclesiastical realm.[146] This causal relationship is, however, difficult to prove.

It is possible, nevertheless, that in the process of the consolidation of canonical doctrine on the subject of the *irregularitates* that prevented a person from taking religious vows, a new concern that affected hermaphrodites especially may have been present. It had become important in the climate of the Counter-Reformation to respond to and dispel the critique of Protestant sects and as part of this attempt to eliminate any suspicion of libidinous behaviour on behalf of the Catholic clergy. As hermaphroditism was on the borders of sodomitical behaviour, clarification on this subject was all the more pressing.

The text that would provide the basis for thought on this subject was the *Tractatus de Irregularitate* (1585) by Simone Maioli, the bishop of Vulturara and Montecorvino. This text systematized all canonical knowledge on the matter that had been accumulated during the second half of the sixteenth century. The key concept, 'irregularity', set out certain physical and other impediments for aspiring clergymen. Different *summae* from the 1500s and, in particular, the work of Martín de Azpilcueta had discussed this question.[147] Illnesses and deformities that impeded the celebration of mass in the required manner or which undermined the dignity of the Church in some way restricted access to the priesthood.[148]

Perhaps unsurprisingly, the condition of the hermaphrodite was named among these irregularities. Irrespective of whether hermaphroditism was deemed possible, following Hippocrates, or unlikely or impossible, following Aristotle, hermaphroditism became a serious problem for the Church authorities and for the individual concerned. Martín de Azpilcueta considered the subject in the Latin version of his *Manual de confesores y penitentes* (first edition, 1549).[149] In a brief paragraph, Martín de Azpilcueta notes that the hermaphrodite should not be ordained if s/he is more female than male and neither should s/he be on being more male than female, as the individual concerned was an object of ridicule (*ludibrium*) as well as being a monster of Nature (*monstrum*). Only the pope could provide a dispensation for such a person.[150]

Later, in works that were published posthumously, such as *Consilia et Responsa* (1591) and *Nunc primum in unum quasi corpus argumentati* (1591), Azpilcueta elaborated upon his previous work. The hermaphrodite who was potent in both sexes could in no way constitute a subject to be ordained. It was not a question of any physical irregularity in this case but rather a moral disorder. If the doubly potent hermaphrodite was ordained, s/he could not take mass as he was not a full male and would, in any case, present a danger for the ordered lives of monks and nuns. Such a presence would be 'indecora et schandalosa' and would contribute to incite the 'libidine' of fellow monks or nuns. Monstrosity was no longer the issue; scandal of an erotic variety was perceived as the threat for life in the convent.[151]

In another passage in *Consilia et Responsa*,[152] repeated in the *Nunc Primum*,[153] Azpilcueta broached a further conundrum for his contemporaries. What should be done if the hermaphrodite, ignorant of his or her status as a dual sex person, was ordained? If this irregularity was discovered at a later time, as had been the case of several nuns who had become men, would the ordination continue to be valid? Could the priest continue in his role with a papal dispensation?[154]

Martín de Azpilcueta was particularly strict on this point: the ordination should be annulled, and the hermaphrodite in question would have to suspend their ministry while awaiting papal dispensation. Other authors, such as Pierre de la Palade, appeared to be more flexible on this point. If the male sex predominated in the hermaphrodite, ordination was not allowed. If ordination had already taken place, it was necessary to accept it as a *fait accompli*.[155] Other Iberian commentators, such as the Franciscans Miguel Rodríguez and Enrique de Villalobos, confirmed the position of Azpilcueta; if 'los hermafroditas son irregulares' (the hermaphrodites are irregular), they would pronounce them incapable of becoming priests even in the case of male hermaphrodites.[156] Rodríguez, however, like Azpilcueta, admitted the possibility of a papal dispensation.

It is therefore possible that the supposed papal dispensation that Donoso claimed he possessed did in fact allow him to remain as a priest in spite of his condition as a hermaphrodite. Given his apparently predominantly feminine aspect, however, it is perhaps more difficult to believe that this was indeed so. What does appear to be certain is that the priest from Zafra, when he was ordained, possessed a male member, albeit of reduced dimensions. It is possible that his being from an influential family, a status used to intimidate witnesses,[157] may also have acted in his favour when seeking a papal dispensation. By contrast, in the cases of the nuns from Santo Domingo del Real in Madrid (around the mid-sixteenth century) and from the Coronada de Úbeda (*c.* 1617), who became men, no papal privilege was mentioned. The first of these was accepted as a monk after her transformation, and she took the name Rodrigo Montes. The second nun, from Úbeda, left the convent and was joyously received by her father, who was delighted to have found an heir. The 'clériga', as he was known in Zafra, used the papal dispensation, whether genuine or not, at his will in order to justify his amorous activities with young men in the locality and did not therefore place in jeopardy his own heterosexual image.

Our discussion of the life stories of Estebanía de Valdaracete, Elena de Céspedes, Catalina de Erauso and Juan Díaz Donoso has allowed us to undertake an examination of several different subjective worlds belonging to the old regime in Spain and in the Americas, from the position of a master at fencing to that of a village priest, from the military adventurers of the Indies through to the marginal world of *morisco* barbers and surgeons. In all these cases, the degree of agency of the individuals studied and their ability to subvert the sexual norms of the era have been highlighted and placed within the internal logics of the social conditions and sex/gender apparatus that were clearly different from our own today.

3 THE EXPULSION OF THE MARVELLOUS: THE DECLINE OF THE 'ONE-SEX' MODEL, 1750–1830

An Unusual Case: Fernanda Fernández, the Capuchine Nun

As we have argued in previous chapters, stories about people who suddenly changed sex were relatively common in literature that depicted the marvels and wonders of the world, the 'relaciones de sucesos' and the anatomical treatises that were published in Spain in the sixteenth and seventeenth centuries. According to the renowned and, by now, questioned thesis offered by Laqueur,[1] sex change and news about the births of hermaphrodites in the human species would concur with the predominance of the one-sex model, which in turn was driven by a medical understanding based on the thought of Galen and Hippocrates. Laqueur also argued that this model began to break down in the West during the Enlightenment and the liberal revolutions and was replaced by a dichotomous model that continues to this day.

However, in Spain at the end of the eighteenth century, in contrast to what was occurring in other European countries,[2] news about sex changes and hermaphrodites continued to be common currency in intellectual and popular spheres. An example of the longevity of the older model is the case of Fernanda Fernández.[3] Fernández was born in Baza and became a nun in the Capuchine convent in Granada. She remained part of the order until the age of twenty-seven, when she started to notice signs of masculinity in her body. In two years the sexual transformation was complete. Doctors initially diagnosed madness, as Fernández explicitly recounted her desires for the other nuns in the convent and her attempts to resist such temptations. Fernández tried to quell these desires by avoiding her companions and by means of strict penitence, including spiked chains and sharpened crosses. Later came regular bloodlettings as prescribed by the doctors. Subsequently, after reiterating her manly nature before the assembled doctors, she was finally examined by them and they pronounced her to be a man. As a result of this revelation, the religious authorities began to process her release from her vows. Once the parents were told, 'Fernanda' became 'Fernando' and took male attire. This new identity was not assumed easily by Fernández, and he pined for his life as a nun and all he had learned in the convent.

But what is most surprising about Fernández's case is that it takes place in 1792 and that everyone involved, including the medical authorities, did not appear to doubt the possibility of sexual transformation, something generally accepted in many other European medical circles to be impossible – the case of Jean Grandjean in France (1765) illustrates this point. The doctors' intervention was limited to reporting Fernández's life experience on the basis of an anatomical examination. In no sense did they argue that deep down Fernández had always been a man or that maleness had always been her true biological sex. However, by this time in both Spain and wider Europe, medical accounts and informed opinion tended to judge this kind of metamorphosis and hermaphroditism as baseless frauds arising out of superstition, a product of the general ignorance of the time.[4] How were such polemics received in the culture of late eighteenth-century Spain? What took place during the nineteenth century that allowed for the rejection of the acceptance of sex changes and hermaphrodites?

The Expulsion of the Marvellous and the Naturalization of the Monster

As the case of Fernanda Fernández shows, the belief in masculinized women and in hermaphrodites was still alive and well in the eighteenth century. The literature on 'marvels' depicting 'strange observations' or 'curiosities' was still very current. The possibility of engaging in carnal activity with the devil was still discussed, and in those cases of conception and birth, the need to baptize the newborn was emphasized.[5] Authors with a certain reputation as 'illustrated' and 'experimental' *avant la lettre*, such as Padre Feijóo, believed in the possibility of procreation between animals and persons.[6] Feijóo also admitted the possibility of bicephalous humans capable of surviving several years and capable of dialogue between the two heads.[7] In sum, in the same way as the political, economic and social *Ancien Régime* was able to survive beyond its supposed disappearance, as Arno J. Mayer has shown,[8] so could the sexual *Ancien Régime* continue beyond its usually accepted demise.

Despite this, it would be incorrect not to admit that throughout the eighteenth century in Spain and in wider Europe a broad and increasingly vociferous offensive against the marvellous took place. This discourse found that true hermaphrodites did not really exist, decried the possibility that they procreated, and refused to accept that women could change into men. In order to explain how this step took place, we need to identify three connected processes: the naturalization of the monster, the development of modern legal medicine and the biological basis of sexual difference. The bedrock of this change was provided by the continual discrediting of the transcendent order that held Nature to be an expression of divine will. Nature began to be understood as mere nature;

life emerged as 'bare life' and as a process governed exclusively by its own laws.[9] Once the protective shield of Providence disappeared, life and nature became fragile, unprotected and dangerous. Their care and protection came to be a political issue. Government would consist above all in directing life, administrating its flows and managing its risks. It would involve, once the old regime decayed, the rise of divisions and taxonomies of human beings.[10]

If Nature is just nature, the monster, which is one of its products, can no longer be viewed either as a sign of divine omnipotence or as a warning or punishment delivered from on high by Providence. During the eighteenth century the process of the naturalization of the monster starts. This process culminates in the first half of the following century with the scientific explanation of monstrosity. This understanding emerges principally in the writing of Isidore Geoffroy Saint-Hilaire in the field of teratology. The naturalization of the monster constitutes its definitive uncoupling from diabolical intervention, from the aberrations of imagination and from dreams.[11] It signifies the conversion of the monster into something housed in the natural order according to laws discovered by reason.[12] This process is an epistemological prerequisite for the entering of the monster into the teratological order. In the context of the apparent duality of the sexes, there are only more or less anomalous genital malformations.

In the course of the eighteenth century the monster becomes an object and instrument of investigation. In the monster the keys to finding the truth of the 'normal' conformation of human beings were located. The monster, in this sense, was utilized in order to carry out experiments in order to resolve the debates between the rival systems of preformationism and epigenesis.[13] It also served, with respect to the first system, to decide between ovism and animalculism.[14] The analysis of the monster also served to resolve the question of the circulatory system of the foetus[15] and allowed eighteenth-century naturalists to delve into the laws governing animal species. How could identities, continual transitions, differences and variations in the hierarchical order of animals be explained? From Leibniz to Robinet, the variations represented by monsters would be conceived either as transitional forms between different species (as a guarantee of continuity) or as a sign of the infinite combinations that nature provided (as a source of difference).[16] Finally, the unification of embryology, which emerged from the triumph of epigenetic theses (Meckel), with comparative anatomy, removed from the idea of an 'animal series' (Cuvier), would give rise in the first third of the nineteenth century to teratology under Saint-Hilaire. Here, monstrosity is placed among the various types of functional anomaly and is placed within the evolutionary register as 'arrested development'.

The extremely frequent observations on monsters in the publications of European science academies between the final years of the seventeenth century and the first decade of the eighteenth century suffer a drastic reduction from

1710 onwards.[17] Strict criteria are enforced not to reiterate case studies and in order to select pertinent cases that explain concrete problems. Above all, there was an attempt to divorce monstrosity from anything to do with admiration of the 'marvellous'.[18] In Spain, the decline of the literature of marvels paralleled the rise in textual critique and analysis,[19] in particular in the mode provided by Feijóo and Mayans.[20] The objective of these studies was to differentiate fables and legends from authentic demonstrable historical facts.

In this context, sexual transformations are steadily deemed to be impossible and, to a lesser degree, so too are hermaphrodites, although they are still the subject of profound controversy in the eighteenth century.[21] Spanish anatomists of certain prestige such as Martín Martínez,[22] religious figures with a naturalist bent such as Hervás y Panduro (1735–1809)[23] and Barco y Gasca (fl. 1775),[24] and figures in 'legal surgery' (a subject included in the curriculum of the College of Surgeons from 1780) such as Juan Fernández del Valle (fl. 1790)[25] rejected the possibility of 'mudas de sexo' (sex changes). Fernández del Valle and Hervás y Panduro also rejected the existence of 'true hermaphrodites'.[26] What is not clear is whether Feijóo actually rejected the existence of hermaphrodites, as he did not discuss the matter directly. Despite this, in a discussion of two-headed monsters he included those who possessed two different sexes: 'unos [de los monstruos bicípites] tenían el órgano de la generación duplicado, otros no; y entre los que le tenían duplicado, en unos le había de ambos sexos, en otros de uno sólo' (some [of the bicephalous monsters] had dual organs of generation and other did not; among those that did possess double organs, there were those that had two sexes and those that had only one).[27]

Martín Martínez, in his *Anatomía Completa del Hombre* (1728), admitted the existence of true hermaphrodites: 'si por alguna contingencia ... quedan colocadas, más o menos partes de las que debían, y mejor o peor elaboradas, sale el fetus monstruoso ... ansí como si los genitales de ambos sexos hallan oportuno lugar de colocación en el debido sitio, puede engendrarse un verdadero hermafrodita, de que hay muchas observaciones, que trae Bonet, contra la opinión de Diemerborch, que no admite hermafroditas verdaderos, sino aparentes' (if for some reason ... there exist more or less parts, whether well developed or otherwise, than there should be, the foetus is monstrous ... just as if the genitalia of both sexes find an opportune place in the correct place, a true hermaphrodite can be engendered. On these there are many observations made by Bonet against the opinion of Diemerborch, who does not admit true hermaphrodites but apparent ones).[28] He does, nevertheless, remind the reader that on more than one occasion women with a prolapsed uterus have been confused with hermaphrodites.[29]

Over the last years of the eighteenth century, the Jesuit and polygraph Lorenzo Hervás y Panduro, in his monumental *Historia de la vida del hombre* (1789–99), ridiculed the belief in hermaphrodites, arguing that they were a myth stemming from ancient times and that they had been reinvented in the modern period as an exoticism and supposedly proved by rather 'puerile' means.[30] He acknowledged the fact that legal specialists, theologians and scientists had accepted their existence and that countries had even legislated on them, and that this may seem to presuppose that hermaphrodites did indeed exist. However, as he stated, 'otras leyes más ciertas, que son las de la experiencia en la naturaleza humana, nos obligan a dudar, y aún a negar la existencia del hermafroditismo' (other more certain laws, providing evidence of human nature, lead us to doubt and even to deny the existence of hermaphroditism).[31]

Hervás y Panduro had recourse to the authority of certain medical figures from the early years of the seventeenth century, such as Riolan and Schenck, to back up his case. These figures, the one French and the other German, employed Aristotelian arguments to combat belief in the existence of hermaphrodites. Hervás y Panduro also drew on the work of the Fellow of the Royal Society, James Parsons.[32] The Spaniard pointed out that the fact that hermaphroditism existed in many plant and animal species should not imply that the same was the case for humans.[33] What in reality occurred was that due to certain 'excrecencias carnosas' (carnal excrescences) in some newborns, the sex of the individual was unclear. He argued that these 'excesses' were more common among women.[34] Hervás y Panduro in this way shared the common medical opinion that many so-called hermaphrodites were in fact 'tribades', that is, women with unusually large clitorises. In doubtful cases he advised certain measures be taken, for example the dressing of the infant in a kind of overshirt or tunic ('hábito talar') until the day when 'señales claras de un sexo determinado' (clear signals of a particular sex) became apparent. The Jesuit believed that come puberty, the true nature of the child would be made visible, and for this reason he suggested that the voice and 'inclinations' of the child be taken as evidence either way.[35]

This would appear to suggest that the thesis rejecting the existence of hermaphrodites, as expounded in Parsons's *Mechanical and Critical Inquiry into the Nature of Hermaphrodites* (1741), was accepted at least in some circles in Spain.[36] The copy of this volume consulted for this research is to be found, no less, in the library of the Real Colegio de Cirugía de San Carlos, Madrid. This institution was established by Carlos III in 1780 as a result of a petition by Antonio Gimbernat y Arbós (1734–1816), who would become the college's first director.[37] It is possible that this eminent Catalan anatomist, trained by John Hunter, another Fellow of the Royal Society, and the founder of the College of Surgery of Barcelona,[38] helped in the dissemination of the thought of Parsons in the new field of legal medicine and surgery that was beginning to be taught from 1795 in these centres.[39]

Juan Fernández del Valle was among the first cohort of surgeons trained at the Real Colegio de San Carlos.[40] Subsequently, he became a teaching assistant in surgery at the general hospital in Madrid, and he would eventually become a permanent surgeon at this same hospital and in the Hospital de la Pasión, also in Madrid.[41] Between 1796 and 1797 his *Cirugía Forense general y particular* was published. In the first volume of this work, which, together with the book by Plenck, would become the main reference text for Spanish surgery schools, the author described the belief in hermaphroditism in humans as a vulgar or popular credence not fit for science.[42] In a section on the male genitalia, he admitted the possible birth of individuals who did not possess these external organs; they would constitute cases of 'monstrosity' which popular accounts would deem 'hermaphrodites'.[43] In a section on female parts, it was also admitted that if these were oversized, they could resemble the male organs in respect of their 'figura, estructura, sustancia y demás requisitos' (figuration, structure, substance and other qualities). Such similarities could well induce 'a los poco prácticos, que había hermafroditas' (those of a less practical bent [to think] that there were hermaphrodites).[44] Once more, the association in medicine between women with large clitorises and hermaphroditism was repeated.[45] In order to dispel any confusion, the manual by Fernández del Valle proceeded to catalogue the differences between the clitoris and the penis and emphasized the incommensurability between female and male genitalia.[46]

The persistence of the 'fábula de los hermafroditas' (fable of the hermaphrodite) was due to an overemphasis on theoretical rather than practical training among medical professionals. Fernández del Valle alluded to 'anatómicos de bufete' (charlatan anatomists), the 'poca o ninguna práctica de la Anatomía' (little or no practice of anatomy) and the 'facilidad que hay en creer lo raro' (ease whereby the strange is believed in) in order to explain the belief in such a myth, which would in turn be reinforced by canon and civil law that offered 'abundantes testimonios' (abundant examples). It was necessary, therefore, to focus on experience to guide the doctor's understanding: 'no hay un solo Anatómico práctico y docto de ninguna época, ni nación, que afirme haber visto ni disecado un solo hermafrodita; lo más que dicen es que han visto monstruosidades' (there is no practical and expert anatomist in any period or in any nation who affirms that he has either seen or dissected a single hermaphrodite; at most they say they have seen monstrosities).[47]

In the second volume of his manual, Fernández del Valle was even more vociferous in his rejection of hermaphroditism. He asked that 'que se borren en nuestros escritos las descripciones de los *Andróginos*' (descriptions of *Androgynes* be struck from our accounts) and demanded that the legal ceremony whereby election of sex and the vow to retain this sex by 'perfect hermaphrodites' was abolished.[48] The rather lurid 'historietas' (storytelling) on the topic of sex change had grown out of this fascination for hermaphrodites.[49]

Such a rejection of hermaphroditism was, however, not yet the intellectual norm of the eighteenth century. This is revealed by a study of one of the most used manuals in surgeons' colleges in the century, especially in that of San Carlos.[50] The *Elementa Medicinae et Chirugiae Forensis* written by the Austrian surgeon Joseph Jacobo Plenck (1733–1807) was published in Spanish for the first time in 1796. This text, critiqued by Fernández del Valle as a result of its belief in human hermaphroditism,[51] still belongs, in many respects,[52] to a dynasty of pre-modern medico-legal works such as the *Quaestiones Medico-Legales* (1621–35) by the Italian Paolo Zacchia (1584–1659) that form the early bedrock of the discipline.[53] The task of the medico-legal doctor in this context was still not divorced from religious elements, and much medical jurisprudence referred to canon law. It is in this framework that Plenck includes his analysis of matters pertaining to monsters, doubts over sex and possession by the devil.

In the section on the signs that might call into question the sex of the individual, Plenck examines the matter of hermaphroditism. He understands hermaphroditism as part of a fivefold problematic: the name with which the child is to be baptized; the legitimacy of the marriage, understood as only possible between man and woman; the determination of the sex of the spouses if both are hermaphrodites; licence to take male or female occupations; and the kind of dress an individual should take up.[54] Plenck considers hermaphroditism as a type of monstrosity that affects the genitalia, giving rise to part-male and part-female forms.

In his account three types of hermaphrodites are described, with their different anatomical and physiological characteristics and their own secondary characters and sexual inclinations. The first type is the 'male androgyne' variety or masculine hermaphrodite. This type possesses a penis and testicles, is capable of inseminating and has an opening in the perineum that seems to be a vulva. On examination, this opening is found not to lead to a uterus but to the bladder. These hermaphrodites are attracted to women, have hair in abundance and beards but no breasts. Finally, they possess a narrower femur and a slightly broader humerus.[55]

The second type is the 'female androgyne' or feminine hermaphrodite. This class possesses a large clitoris which appears to be a penis and is capable of erection. This hermaphrodite usually possesses two openings, one of which leads to the bladder and the other to the uterus. They have no testicles or spermatic ducts. They possess breasts, scant body hair, a broader femur and a narrower humerus.[56]

For Plenck there is a third type of hermaphrodite, the 'true hermaphrodite'. These possess a mixture of the sexes and have testicles and ovaries, a uterus and virile member. In order to demonstrate that this variety actually exists, Plenck cites observations and cases collected in the works of Haller, in accounts by the French authors Mavret and Petit – read at the Dijon Academy and the Royal Academy of Science, respectively – and from the work of the Italian Colombo.[57]

His chapter ends by formulating five theses. These are: a) male androgynes can inseminate women; b) female androgynes, using their clitorises, can unite with women but cannot ejaculate; c) true hermaphrodites are possible;[58] d) the existence of these true hermaphrodites explains stories about women changed into men and vice versa, yet what in fact takes place is that the genitalia of the other sex emerge from inside, either as a result of an operation or on coming of age; and e) old laws punished these unfortunates severely, and Nature has punished them sufficiently as it is.[59]

Plenck's account can be understood as heralding a transition.[60] On the one hand he maintains the belief in true hermaphrodites, and on the other he denies the possibility of sex change. But in order to deny this possibility, he uses an old device: sex changes arise from hidden hermaphroditism. Finally, he adds an element that is present in the intellectual armoury of the Enlightenment and that Spanish medico-legal doctors will reiterate in the first half of the nineteenth century – he condemns the old laws as barbarous hangovers, which punish savagely individuals of doubtful sex, and displays his sorrow for these 'unfortunate beings' that Nature has created with horrendous deformities.

The Medico-Legal Doctor as Ultimate Authority on Sexual Identity

The work by Plenck forms part of a first generation of texts in legal medicine, which includes *Cirugía Forense* (1783) by Domingo Vidal,[61] *Cirugía Forense, General y Particular* (1797) by Juan Fernández del Valle, and to some degree the *Compendio de Policía Médica* (1803) by Vicente Mitjavila.[62] In these texts, which are used as manuals in the colleges and, in the last case, in the Academy of Practical Medicine in Barcelona,[63] legal medicine goes far beyond the parameters of the discipline as laid out at the time of Paolo Zacchia. It is no longer a simple discipline that tries to guide justice in particular circumstances such as violent deaths, poisoning, witchcraft, rape, etc.; it is now a science of the state.

We stated that the naturalization of the monster was only possible by means of the wearing away of the concept of Nature as a language through which God communicated with humanity. The role of a transcendent order that gave sense and protection to life is gradually supplanted by a disciplinary form of power characteristic of absolute monarchies and which attempts to administer the tiny everyday details of life. The 'science of police', both in its French and German varieties, is the theoretical model followed by this type of power.[64] The German variety, which was disseminated in Spain in the eighteenth century, constituted a set of knowledges devoted to the creation of state administrative functionaries, also known as 'cameralism'.[65]

Part of this set of knowledge was 'medical policy'.[66] If the management of life and health were integral to a state that watched over public well-being, it is not surprising that public health was recognized as an important field, particularly in the reign of Carlos III.[67] In fact, the first series of medico-legal texts mentioned above were to some degree inscribed in the framework provided by 'medical policy'. If the state and its laws should protect life, this was only possible if the principles that guided it were known. This is evidently the case for the question of the 'population', considered at the time to be the major form of wealth of the nation.[68] Legal doctors, therefore, as major specialists in health and public well-being (at this time, public hygiene was not separated from legal medicine),[69] were not limited to guiding magistrates. Their knowledge should be employed in the close regulation of life in order to guarantee the harmonious functioning of the state.[70]

This idea of subordinating law to biological norms is articulated in these first texts in an extraordinarily centralist and interventionist mode, corresponding to the mechanisms of state characteristic of enlightened despotism.[71] Prussian medical policy,[72] drawing on an older juridical tradition stemming from Lutheran reform, did not permit the hermaphrodite to elect his/her sex even in doubtful cases. It was the doctor, as an agent of the state, who should establish the civil sex of the individual.[73] This position, which contrasted with that of the Spanish theologians of the Counter-Reformation who allowed for the conscience of the affected individual to take precedence over the criteria of the experts, would remain at the heart of the Prussian state and would be explicitly written into a law on hermaphrodites in the nineteenth century.[74]

Spain would be different. The emergence of the liberal state, which in Spain, after the failed attempts of 1812 and 1820, is resuscitated under Isabel II, implied a very different way of implementing the government of health and the precepts of legal medicine. In this new context, the legal doctor does not collaborate with a state that wishes to regulate life meticulously as under cameralism and the days of medical policy. The liberal government of life consists in eliminating the obstacles that prevent the development of the means of internal regulation of life; this form of government marks out the limits and the possibilities of state action. It is not a matter of submitting nature to an endless round of sanitary interventions. Rather, the very dynamic of vital processes is submitted to analysis so that the legislator can adjust his actions accordingly.[75] From the beginning of the nineteenth century, with the Spanish translation of the work of Foderé, *Les Lois éclairées par les sciences physiques, ou Traité de médecine légale et d'hygiène publique* (1797), published in Madrid between 1801 and 1803, a second generation of medico-legal treatises is opened up. The work of Ramón López Mateos (1771–1814), *Pensamientos sobre la razón de las leyes* (1810), and Francisco Fabra Soldevilla (1778–1839), *Filosofía de la Legislación Natural* (1830), inaugurate a corpus of Spanish medico-legal texts[76] that are faithful to the model of liberal governmentality, whereby legislators should adjust their work 'a las insinuaciones de la naturaleza'.[77]

Among those areas brought under the aegis of legal medicine, there is one that is particularly relevant to the achievement of the optimum quantity and quality of the population. The medico-legal practitioner became the ultimate authority in respect of the assignation of sex to those individuals deemed of 'doubtful sex'. No longer is the 'predominant sex' spoken of or the election of sex in cases of 'true hermaphroditism'. It is understood that each individual has one exclusive sex, male or female, which guarantees their civil identity and holds up the institution of marriage, the key mechanism for the reproduction of the nation.[78] The legitimate procreative couple is the married couple, and such a formula requires the strict identification of a man and a woman.[79]

The desire to regulate marriage beyond the sphere of family interests alone in order to create an abundant healthy population is what engendered in educated Spaniards a critique of marriages of convenience and marriages with large age differences. These matrimonial alliances were unsuitable not only morally but also because they affected the biological potency of the realm. Such concerns can be seen in a wide range of sources, including comedies of the time such as that of Leandro Fernández de Moratín, *El Sí de las Niñas* and *El Viejo y la Niña*, or the satires of Jovellanos on the same matter.[80] In the first of these pieces, from 1786, marriages of convenience, adultery and libertinage, declared as especially common among the nobility, were subjected to ridicule. High-class women, drawn to marriage on the promise of luxurious wealth,[81] submitted themselves to the 'vejez hedionda' (putrefying old age) of the country,[82] and thus encapsulated the decadence of the times and their contrast with the older morality represented by the upright 'Matronas castellanas' (Castilian matrons).[83]

Goya's *Caprichos*, drawn between 1797 and 1798, also fall into this camp. They illustrate in grotesque form a kind of marriage ceremony in which the protagonists are masked figures in a kind of carnivalesque scene. The fiancée, a young girl, appears with a mask on her face and another on her genitalia, simulating the double sex of the hermaphrodite. One of the illustrations of the series (B. 59) is entitled *Máscaras. La apunta por hermafrodita.* A scribe, in likeness to the priest, holds a notebook in which he appears to record the proceedings. Behind, someone watches with hands raised in horror. The scene has been interpreted as an allusion to female luxuriousness common to the *novias* in this kind of marriage, which, by agreement with the boy, is arranged in order to satisfy the girl's unruly erotic appetite.[84] Hermaphroditism is symbolized by sexual excess and draws on the old association with nefarious sin. But at the same time, this hermaphroditism is not real. It does not coincide with that represented in the literature of prodigies in contemporary accounts and is not represented as a harbinger of evil to come. It is no more than a mask, an appearance that hides the true sex of the individual. In this way, Goya's illustration appears to draw on the hermaphrodite as a transitory condition and at the same time suggests that the old sexual regime is fading away as the reign of the true biological sex approaches.

Such a depiction implies the uncoupling of sexual identities from the old network of community and family alliances that characterized the old regime. Here, the civil identity of the subject was defined by his or her external lines of sociability. It was necessary to determine clearly the sex of a person in order to allow their entry into relations governed by alliances, their entry into the ecclesiastical order or their participation in marriage. Identity also allowed them to be positioned with respect to lineages and inheritance, in guilds and corporations which required a name and a tradition. Determination of sex in doubtful cases was primarily the responsibility of the family or tutors, who in turn often sought the guidance of doctors, surgeons and midwives. In this way, the individual was defined less by their sex than by their relations with others. Possessing one sex or the other determined whether the subject would participate in a dense set of relations of dependency (family, vows of fidelity, protection) as part of the social network provided by family and blood alliances.

The new liberal state, which eliminates the representation of society as divided into three unchangeable orders and substitutes a homogenized society in accordance with property relations, defines the social identity of individuals not by their names or titles or their external relations but by means of their 'interiority': their body, their physical strength and their thought, all elements that were identified with the responsible individual who was capable of exercising their rights and entering in the contractual relationship.

The fixing of sexual identity, replacing status as a distinctive and innate mark in individuals, would no longer depend on the members of the family or on the subject themselves. Even though the subject would be invited to speak the truth about him- or herself, this task falls on those who possess positive knowledge on bodies and souls and who are capable of deciphering definitions beyond any deformities that nature might throw up and beyond any interpretations that superstitious priests might put forward.[85] This is a technical form of rationality. Its social agents possess expert knowledge which even supersedes any juridical authority in terms of determining the identity of the subject.

These new forms of administrative rationality are devoted to seeking the maximum output from the combined strengths of the nation. The science of administration of the old regime and the political economy of liberalism both see the population as a source of wealth, a resource that should be managed to full effect and a treasure whose increase, both quantitatively and qualitatively, impacts on military might and the productivity of the state.

The establishment of a national militia under a system of conscription and the concern with the regulation of the age and ability of marital arrangements to reproduce are measures wholly in line with this desire to increase the volume and health of populations.[86] In both scenarios, the identification of the true sex of the hermaphrodite has become the principal objective. This would engender not only happiness and fulfilment in marriage but also the maximization of genetic possibilities whereby the subject would be authorized, or not, to engage in marriage or to have their marriage annulled should the need arise. As factors to be taken into account for these decisions, problems of impotency, sterility, age, health, hereditary diseases and the assignation of sex would all become the remit of the new medico-legal sciences.[87] As an example of this shift, the diagnosis of impotence as a cause in the annulment of marriage can be cited. No longer is public proof of the couple's cohabitation required.[88] The matter has been privatized, and certification is made on the basis of an interview of the couple and the clinic examination of the body, to be undertaken by the medical expert.[89]

The identification of true sex would also be required as a prerequisite for military service, which from 1770 was undertaken on the basis of the 'sorteo' and from 1830 using the 'quintas' system.[90] Hermaphroditism, alongside the lack of a penis, syphilis, gout, cancer of the eyes, alopecia and a host of other conditions, would provide the basis of exemption from military service.[91] It comes as no surprise to note that many cases of supposed hermaphroditism attended to by nineteenth-century legal medicine were to be found among fiancés and those married, or in soldiers of 'ambiguous' bodily features.

The Foundation of Sexual Difference

Together with the take-off of legal medicine and the naturalization of the monster, there is a third process during the same period that is decisive in terms of the eclipse of hermaphrodites and the rejection of the possibility of sex changes. This third element is the biological foundation of sexual difference, a development that, as Thomas Laqueur has argued, converges with the aspirations of enlightened thought and liberal democracy. The route taken by such a development is now traced here for the Spanish case.

Emphasis on the duality of the sexes, contrary to the monist schema of Hippocratic-Galenic thought, can be traced in some medical texts (Bravo de Sobremonte, García Carrero) and some non-medical texts (Martín de Río) of the seventeenth century.[92] But in these kinds of texts, divine will is always invoked: two sexes exist because God wished it so. The Book of Genesis proves it, and this guarantees the reproduction of humankind. This theological argument characteristic of the *Ancien Régime* will lose ground in the light of new interpretations. It is not necessary to read the Scriptures to see differences between men and women; it is a case of deciphering different traits in anatomical structures, physiology and temperaments. It is from this perspective, although without renouncing the old theological and moral interpretations completely, that Feijóo and Martín Martínez argue.[93]

In the *Teatro Crítico Universal* by Feijóo, the author is extremely critical of the Hippocratic doctrine.[94] In matters relating to generation, Feijóo maintains pre-formationist theories and rejects the notion that female foetuses lie on the left and males on the right.[95] But it is not in these differences with Hippocrates that we see the Benedictine's thought on the differences between the sexes. This can be seen in his 'Defensa de las Mujeres' within the *Teatro Crítico Universal*. This text, as Mónica Bolufer has shown, puts an end to the debate begun in the medieval period in Spain on the inferiority or otherwise of women with respect to men. In his 'Defence of Women', Feijóo follows an argument sketched out in the eighteenth century and refuses to accept the Aristotelian notion of women as 'monsters' or as 'imperfect males'.[96] Instead, women are understood as complete forms, perfect in their own right and biologically necessary 'pues no puede conservarse la especie sin la concurrencia de ambos sexos' (because the species cannot be conserved without the coexistence of the two sexes). As a consequence of this position, Feijóo rejects the theological supposition which, as we can be seen in Eiximenis, foresaw the conversion of all women into men come the Resurrection.[97]

In addition, Feijóo also recognized that men and women were of 'diferente organización' (different organization) and that this physical difference conditioned the moral and intellectual orders. But he insisted that it was not from this difference that any intellectual inferiority of women sprung. Indeed, given the state of science at the time, it was not possible to identify the material basis of any such hierarchy. In this sense, it was necessary to consider both sexes as equal.[98] The difference between men and women was not on the basis of understanding, even though he recognized that women's brains were made up of softer fibres. Such a structure, however, did not undermine her 'facultad discursiva' (discursive ability), as 'illustrious' and intelligent women throughout history showed.[99] The difference between the sexes lay elsewhere: not 'en los órganos que sirven a la facultad discursiva; sí sólo en aquellos que destinó la naturaleza a la propagación de la especie' (in the organs that serve discursive ability, but only in those in which nature placed responsibility for the reproduction of the species).[100]

Therefore, purely physical considerations did not permit the hierarchical model as conceived by Aristotle (the woman as a failed man) or by the Hippocratic-Galenic model. Women and men were different by their nature, although this did not allow for the supremacy of one sex over the other. To this degree, it would appear that Padre Feijóo anticipated all the aspects of the dimorphous and naturalist model. But this 'modernity' as evinced by a reading of Feijóo is somewhat precarious. If both sexes were equal in terms of talent but different in terms of physical traits, how could the *de facto* supremacy of one sex over the other be accounted for or justified? In order to resolve this conundrum, Feijóo had recourse to a theological argument. God had wanted man to exercise power over woman not because of his superior intelligence but because of other virtues that were in the possession of those in positions of command: constancy and fortitude.[101]

This reference to 'constancy' as an emblem of virility would enjoy a healthy future. As part of a recovery of Stoicism, enlightened Spanish thought that was closest to political republicanism (including the thought of Feijóo)[102] identified this feature as part of an individual's capacity for self-control and dependency on one's own resources in the face of adversity.[103] A particular genre of literature from the early 1800s contrasted the expression of virility held dear by peoples who defended their liberty with the effeminacy of the tyrant who threatened the age-old freedom of the population. Orientalist representations of the 'sexual monster' were thus harnessed to the rejection of a particular form of erotic decadence.[104] Such an association permitted the comparison between the Moorish invasion and the incursion of the Napoleonic troops in the early nineteenth century.[105] It was this same reference to 'constancia' that, in the discussions that led to the Constitution of 1812, allowed for the exclusion of women and Spaniards 'nacidos en África' (born in Africa) from enjoying political rights.[106] Women and the 'racially other' were deemed not to behold the necessary qualities for their own self-management, a capacity enjoyed by the virile Hispanic male.[107] This model of masculinity and constancy was also contrasted with the model of the war-like conquistador and knight that we saw operating in the case of Catalina de Erauso. This old stereotype not only had not disappeared but had also been invoked by those who defended a form of nationalism that was not republican in essence but ethnic.

This political expression close to republicanism identified the nation as an example of masculine constancy with a historical community that enjoyed age-old freedoms and rights as provided by traditional institutions such as the councils, courts and 'fueros'. On an anthropological and moral level, the old, ethnic prototype of virility was satirized by Jovellanos in 1787 in his piece 'Sobre la mala educación de la nobleza'.[108] Here, Jovellanos on the one hand ironized the traditional and 'castizo' fashion of the 'majo' which a section of the aristocracy had adopted. The ignorant, leisurely, brutish but hapless courage of this noble figure, keen to mimic the most ill-advised behaviours of the common people in respect of their love of bullfights and brothels, was rejected unforgivingly.[109] On the other hand, Jovellanos mocks the French-influenced aristocracy, which pours scorn on its own Navarrese, Basque, Cantabrian or Asturian origins and opts to be educated in Paris or at the Sorèze military academy, plagued with vice and temptation.[110] In this case, the target of the author's satire would be 'el alfeñique perfumado y lindo' (the perfumed and beautiful weakling), dedicated to his dress mode and physical adornment, consumed by luxury and libertinage.

The 'putañero', an individual who had lost his 'salud y bienes' (health and wealth) and who before gaining his 'cuarenta abriles' (fortieth year) became old before his time, was also criticized. If this kind of person married, the consequences would be wholly negative: he was 'tímido, exhausto, sin vigor' (timid, exhausted and without vigour) as a result of his lifestyle, he would be condemned to impotence, and his wife would suffer a life of misery.[111] Jovellanos, rather nostalgically, contrasted this de-masculinized individual with the bellicose masculinity of the Spanish conquistadors.[112] But it is important to note that the 'virtud' (virtue) sought by Jovellanos is not a military form of virtue but that of a civic elite that fulfils its responsibilities and provides an example of correct citizenship, having embraced the qualities of hard work, abnegation, culture and frugality.

The ethnic (or ethnicist) interpretation assimilated the nation within a cultural community that was defined in essentialist terms such as character and religiosity. The best eighteenth-century exponent of this particular interpretation is represented by Cadalso in his *Cartas marruecas* (published posthumously in 1789). This author adopted the trope of effeminacy as the backbone of his historical metanarrative. It was luxury and sensuality that brought effeminacy to the Visigoths (an ethnicity that was exalted, in contrast, by the republican tradition),[113] who in turn succumbed to the Moorish invasion.[114] The heroism of the Reconquest – Cadalso, as a professional soldier, praised military prowess – brought a renovated current of virility to the country that culminated in the period of the Catholic monarchs. However, the Spanish people, who had degenerated into effeminacy and were dazzled by the wealth derived from the Americas ('afeminado con el oro y plata de América' (made effeminate with the gold and silver from America)),[115] as well as being exhausted by European war and emigration to the colonies, returned to a state of weakness that would prevail until the eighteenth century. Drawing on Herder, Cadalso considered that luxury, together with racial mixing and the lack of cultural cohesion, cemented by a cultured, peaceful and cosmopolitan aristocracy, had engendered national ruination.[116]

The identification of virility with constancy, an aspect present in the thought of Feijóo and which emerged in republican patriotism, did not imply a completely dichotomous sex/gender model. However, the *Anatomía Completa del Hombre* (1728) by Martín Martínez, an acquaintance of Feijóo[117] and a supporter of the latter's argument in 'Defensa de las Mujeres', shows a clear example of the triumph of sexual dimorphism. In the fourth and fifth lessons, dedicated to analysing male and female parts, differences in these organs are continually foregrounded. The fourth lesson opens with a discussion of the similarities and differences between the male and female organs.[118] In this detailed account, the tiniest differences between these shared parts are emphasized.[119]

When describing the female parts, Martín Martínez emphasizes with great care their function, whether in conception or in birth, as if the whole of female genital anatomy were destined to make the semen fecund and the foetus viable.[120] With regard to the clitoris, although it is compared to the penis, well within the Galenic isomorphic tradition, its differences in terms of structure and function are highlighted. Its muscles are not to maintain an erection or to expel semen but to close up the vulva in order to squeeze the penis during coitus. Also within this tradition, as seen in some medical texts in the sixteenth and seventeenth centuries, is the insistence on characterizing the clitoris as the principal organ of female 'deleite sensual' (sensual pleasure)[121] and the reference to macro-clitoridean women who are capable of seducing and having carnal relations with women.[122] In the work by Martín Martínez, an illness that became fashionable at the end of the eighteenth century is not mentioned. This specifically female disease, of 'furor uterino' (uterine fury) or 'ninfomanía' (nymphomania), was glossed for the first time in the work of Bienville (*La Nimphomanie ou Traité de la Fureur Utérine*, 1771) and was associated with the excitement of the clitoris.[123] This particular pathology, discussed in Spain since the end of the eighteenth century,[124] marks out the differences between women's sexual desire and that of men and thus serves to recapitulate differences between the sexes.

The figures representing the genital organs of man and woman in the work by Martín Martínez, even though they still depict a frontal section, do not show the sexes side by side, as can be seen in sixteenth-century medical texts such as the *Historia de la Composición del Cuerpo Humano* by Valverde de Amusco. The representation of the vagina as a penis has disappeared and the breadth of the uterus and womb is emphasized, forming a conical figure, while the penis is represented as cylindrical. Isomorphism between ovaries and testicles is no longer a given either. In sum, the dichotomous model is now evident.

This break with the one-sex model as represented by authors of the first Spanish Enlightenment such as Feijóo and Martínez will be consolidated as legal medicine becomes established. The 'nueva ortodoxia ilustrada' (new enlightened orthodoxy), as Mónica Bolufer has called it,[125] does not limit these biological differences to the organs of generation; by questioning the old Cartesian dualism, biological difference is understood as the material basis that grounds the mental and physical existence of individuals. As such, doctors such as Foderé[126] and López Mateos[127] do not merely underline the different physical make-up of men and women but also argue that all social differences stem from this different organic constitution. In this way, the universal equality heralded by the collapse of the old regime is questioned, as differences are no longer situated in differences of rank but in the differences between the organic and physiological constitution of bodies. Doctors, like the authors of moralizing novels at the end of the eighteenth century,[128] appeal to 'nature' as against 'artifice' as a way of questioning the lifestyles of privileged groups of the *Ancien Régime*. But this did not eliminate discourse of female inequality; rather, it recodified it and situated this difference on the level of complementary but different biological realities. The social is predicated on biology. Women and men are theoretically equal as juridical subjects, but their physical peculiarities make them more apt to fulfil certain occupations rather than others. The new great divides that characterize industrial society – public/private, production/reproduction, factory/home – are consolidated on a biological basis of bare life rather than in respect of a divine order. At the same time, the sciences of life are invoked to locate the differences between races, ages and classes, thus consolidating a kind of 'state racism' in parallel to democratic liberalism.[129]

This emphasis on the complementarities and differences between female and male natures was consolidated in an emerging set of medical texts in the early nineteenth century. These treatises on 'gynaecopathy' or 'women's illnesses' were represented by works such as that of Julien Joseph Virey (1775–1846), translated into Spanish by the anatomist Manuel Hurtado de Mendoza (*c.* 1780/5–1849). This work emphasized that sex differences were greater the higher one went up the biological scale of living things,[130] and women were declared to be fragile and of acute sensibility.[131] It was this extremely fragile female condition that gave rise to a genre of texts on women in Spain of French origin. There was no male equivalent. This literature – from the treatise by Vigarous, translated in 1807,[132] to that of Baltasar de Viguera (1827),[133] to those of Roussel[134] and Capuron,[135] both translated into Spanish in 1821 – confirmed the biological origins of sex differences and placed these at the root of all female peculiarities, both psychic and social.

The consolidation of sexual dimorphism led to the elimination of any transitional figures on the masculinity/femininity scale. Sexual metamorphoses had become mere fables, and hermaphroditism (apart from 'perfect' hermaphroditism) was increasingly qualified as apparent, although teratology admitted it as a kind of variation within its taxonomies.[136]

The Inheritance of the Enlightenment in Early Nineteenth-Century Medicine

Spanish medical thought of the first decades of the nineteenth century, situated on the confluence of the processes examined above – the naturalization of the monster, the take-off of modern legal medicine and the biological foundation of sexual dimorphism – brought together a wide range of understandings on hermaphroditism and sex change that could be found as part of the intellectual legacy of the Enlightenment.

In the first place, there was the more or less vehement rejection of the biological possibility of true hermaphroditism in humans. Despite not receiving identical treatment across the disciplines of natural history, anatomy and legal medicine (recall that discussions had not been unanimous in the eighteenth century), there was a general rejection of the notion of human beings with two sexes. Increasingly, the belief in hermaphrodites would be consigned to the world of fables, magic and common superstition. Rather, these figures would form part of the fascination with the marvellous, in turn understood as part of humanity's early childhood.[137] In the same way as popular medicine was to be exiled by an increasingly professionalized form of medicine, folk stories about hermaphrodites had to be consigned to history. Hermaphroditism, like sexual metamorphosis, was an example of unreason. Indeed, such notions of the hermaphrodite constituted an error; the very term that was used to designate hermaphroditism, Orfila pointed out,[138] could only lead to falsity, to a fallacious use of language that should be expelled by new medical knowledge. The hermaphrodite came to be defined negatively, as a deceit that nature had fabricated in order to fool positive observation.[139]

Appearance, 'simulacrum', 'deceit' are the terms that became associated with the hermaphrodite. This negation is not just epistemological, a result of the ignorance and backwardness of medicine. The belief in these beings also entailed moral negativity. As Enlightenment thinkers believed – and such is present in the work of Plenck – superstitions were associated with barbarous acts, aberrations of reason. It was because of this absurdity that innocent persons identified as hermaphrodites were executed, in the same way that belief in witchcraft and possession by the devil resulted in the fires of the Inquisition.[140]

This rejection of the hermaphrodite was founded partly on a critical evaluation of past accounts of supposed hermaphrodites. Artistic and literary accounts and supposedly scientific observations made in ancient times made up a mosaic of horrors that reason could only substitute by positive observation in the face of prejudice.[141]

In the same way as in other areas of medical specialism such as the dissection of cadavers in pathological anatomy and the study of mental illness, prejudice and superstition are expelled by the clarity of the clinical eye. Observation, deriving from anatomical inspection, recourse to the microscope[142] and studies of the physiology of reproduction, would constitute the principal basis from which to argue against the existence of the hermaphrodite.

However, these inductive techniques do not create a field informed by complete unanimity. If some authors declared that supposed hermaphrodites observed to date were not capable of reproduction via self-insemination, as occurred in the plant world, other authors pointed out that these beings were not capable of being fecundated or able to procreate. In neither case were they real androgynes. This was the point of view that gained the upper hand.[143]

Alongside these points of view, the occasional author declared that hermaphroditism was impossible *a priori*, a contradiction in terms whether proven or not by inductive means. This argument seemed to go back to old-style understandings. Here, the hermaphrodite would signify the rescinding of 'las leyes que le plugo al Supremo Hacedor establecer en orden a la reproducción de los seres animados' (the laws that our Supreme Maker deigned to establish in order to bring about the reproduction of living beings).[144]

The true hermaphrodite disappears from the stage, to be consigned to the lowest order of living things among plants and inferior animals. At most, he or she would be identified with a certain lack of sexual differentiation characteristic of old age or childhood in humans.[145] Humanity and hermaphroditism are deemed to be mutually exclusive concepts. Only when man is not yet a man or when he begins not being so, assailed by old age and death, is sex erased and a loss of identity similar to that of the androgyne is gained.

In this way, hermaphroditism was completely voided of its magical and occult content. Any case was in reality the result of an anatomical or functional maladjustment that made the assigning of true sex a more complex task. The parameters of this medical discussion have shifted from those of the sixteenth and seventeenth centuries. No longer are questions asked on the 'predominant sex' of the individual. Instead, when faced with an individual of doubtful sex, embryology and teratology would ask: what kind of physical alteration are we presented with? Is this a vice of conformation or a monstrosity? Is this a product of an interruption in the growth of the individual or one of thwarted development? Legal medicine, on the other hand, would ask about the true sex of the individual *hidden behind* the mask of deformity. It would inquire as to the type of apparent hermaphroditism present and would seek an answer to the question as to whether surgical procedures could 'correct' any anomaly to restate the true sex of the individual.

The task of biological discourse is to identify those anomalies that were previously understood as hermaphroditism. By using anatomical and physiological criteria, the teratology of Saint-Hilaire offered a classification of anomalies according to varieties (anomalies that did not interfere with normal functioning), vices of conformation (generally inconsequential anomalies), heterotaxias (severe anomalies but which did not interfere with normal functions) and monstrosities (severe anomalies that did impede normal functioning).[146]

Cases of apparent hermaphroditism were located mainly in the category of vices of conformation and monstrosities. The first of these were not very severe deformities and there were no traces of both sexes to be found. In monstrosities, there was the coexistence of organs of both sexes and this prevented sexual relations and reproduction. Both anomalies were understood as elements in ontogenetic evolution – they were developmental shortcomings resulting from the lack of growth of one type of genitals which had been surpassed in growth by the genitalia of the opposite sex.[147]

In the discourse of teratology, then, hermaphroditism was in reality a kind of underdevelopment in particular humans. The organism was unfinished and was closer to its own origins than to any finishing point. Such a circumstance had echoes on the phylogenetic level. Those species in which hermaphroditism was common were placed on the 'último peldaño' (last gradation), in the words of Pedro Mata, of life, separating plant from animal life, the living from the inert. The hermaphrodite was pure negativity and as such would never become and would never exist.

This notion of limit, of negativity, would be different in the case of legal medicine. Here, as has been argued, hermaphroditism is not seen as being close to any origins but as a figure giving rise to equivocal views. The hermaphrodite is a fiction whose truth is to be unmasked by the medical doctor who locates the real sex. The true sex would be identified firstly as 'true genital anatomy' in the rules established by Henri Marc in 1817,[148] and later by means of the 'gonads' in the histological criteria advanced by the German Theodor Albrecht Edwin Klebs (1834–1913).[149] These new explorations of the 'hermaphrodite' lie beyond the scope of this present work.

4 HERMAPHRODITISM IN PORTUGAL

This chapter locates a number of cases of suspected hermaphroditism in Portugal within the prevailing local framework of sixteenth- to early nineteenth-century interpretations on the question of sexual ambiguity. On doing so, it acknowledges that thought on this subject can be described as an example of transnational knowledge exchange between the kingdoms of Iberia existing at that time and, more broadly, between Spain, Portugal and other European countries, in particular France, Germany and Italy. The desire to pinpoint precisely which ideas came from specifically Portuguese theologians and doctors and which from their Spanish counterparts is a somewhat elusive task given these exchanges. What we seek here, therefore, is not a national history of thought on hermaphroditism. Indeed, the very fact that many commentators on hermaphroditism and sex change – such as Amatus Lusitanus (b. Castelo Branco, 1511–68), Rodrigo de Castro (b. Lisbon 1546), Zacuto Lusitano (b. Lisbon, 1575–1642) and Isaac Cardoso (b. Trancoso, Beira Alta, 1603/4–80) – were either born in Portugal, *marranos* descended from Spanish *conversos*,[1] educated at universities such as Salamanca, or wrote their major contributions to medical thought in cities such as Antwerp or Verona blurs such easy distinctions and stories about any supposed national origin of scientific thought.

The opening section of this chapter focuses on some examples of sexual and 'gender' ambiguity (with the caveats on this terminology already highlighted in the Introduction) in Portugal from the 1500s onwards in order to provide a context for the cases of hermaphroditism that are presented later. In this section, the legacy of gendered transgressions such as that of Saint Uncumber, known also as Saint Wilgefortis and Saint Liberada, the bearded saint whose cult became popular in the fourteenth century and whose life and death were often depicted as having unfolded in Portugal, is mentioned as an important framing discourse for sexual ambiguity in the later centuries. Other cases of sexual transgression, such as that of Antónia Rodrigues, who became António in the early 1600s in order to crew a grain trading ship bound for the Portuguese enclave of Mazagão in what is now Morocco, are analysed for their insights into understandings of the nature of sex (maleness and femaleness) in the period.

In the second part of the chapter, two principal interpretive axes will pertain. The first will analyse the thought of philosophers and medical doctors, many of Jewish origin such as those mentioned above, who framed discussions on sex mutation and hermaphroditism. Their thought will be placed in an international, especially Iberian, frame and will also be set against the backdrop of the changing role of science within the context of the overseas explorations that led to the Portuguese Discoveries. The second interpretive axis will analyse the relationship between the conceptualization of the sins of sodomy and heresy and their connections with possible hermaphroditism in individuals identified as having transgressed social and religious norms, particularly in light of the actions of the Inquisition that operated in Portuguese society. Within this section, a more general cultural analysis will also shed light on popular attitudes to these phenomena.

The third section of the chapter provides an in-depth analysis of a number of cases of hermaphroditism that occurred from 1500 up to the end of the seventeenth century in light of the framework set out in the previous two sections.

Sexual Ambiguity in Early Modern Portugal

The legend of Saint Wilgefortis, known in the English-speaking world as Saint Uncumber, in Spain as Saint Liberada and in Portugal as Saint Librada, is one example among a cluster of cases of 'bearded saints' who changed their appearance in order to avoid an undesired marriage or undesired male attention.[2] Purportedly the daughter of a pagan king of Portugal, she undertook a vow of chastity but was promised to the king of Sicily against her will. She prayed for release from such an agreement and grew a beard to deter her suitor. Once he saw her, he shied away and any interest in the marriage faded. In revenge, her father had her crucified.[3] She thus became the patron saint of women who wished to 'uncumber' themselves from male attention or undesired husbands. Another case in the Iberian Peninsula was María de Gelbes. The Toledo-born María de Gelbes dressed as a man in order to flee from her own forced marriage. In the small village of Puebla del Deán (in La Coruña province), she joined the Franciscan convent. Once she revealed her sex in a confession, she was removed to the Clarisas de Pontevedra convent, where she died aged forty-four.[4]

Of a non-religious nature was the case of Antónia Rodrigues. In his posthumous account of the kingdom of Portugal, its geography, plants, natural resources, saints, noblemen and prominent women, the royal chronicler Duarte Nunez do Leão (fl. 1530–1608) told the story of a fisherman's daughter from the town of Aveiro who took up men's clothes, cut her hair in the masculine style and crewed on a grain ship bound for the coast of North Africa in the late 1500s. The story of Antónia Rodrigues, whom Nunez do Leão met during his lifetime on her return from Africa to Lisbon, was recounted as evidence of one of the many feats achieved by Portuguese women in the history of the country.[5] The life of Antónia Rodrigues, retold and referred to as an inspiration for poems and short stories well into the twentieth century,[6] although not one of hermaphroditism, comprised cross-dressing and the inevitable possibility – eventually deferred – of same-sex passion. Rodrigues, daughter of Simão Rodriguez and Lianor Diaz, came to Lisbon at the age of twelve with her mother to sell clothes. She took to trying on male attire and settled on a vestment 'conforme ao trajo dos moços que seruem no mar em nauios merchantes' (appropriate to the dress of young men who served on merchant ships on the high seas).[7] After cutting her hair short, she went to the docks and parlayed with the captain of a ship that was to take wheat to Mazagão. Once at sea, Rodrigues acted as a crew member 'como se fora homem que fizera sempre aquelle oficio' (as if she were a man who had always undertaken this office) and just like a 'destro marinheiro' (qualified seaman). Early on in the voyage, she decided to change her given name from 'Antónia' to 'António'.

On arriving safely at Mazagão, questions were raised by the port authorities in respect of the legitimacy of the cargo, and the 'grumete' (lad) António Rodrigues candidly gave his views highlighting certain irregularities. As a consequence of the resultant ire of the captain, he feared for his personal safety. In order to escape any vengeance on the captain's part, Rodrigues enrolled as a soldier in Mazagão fort and battled with the Moors in order to maintain the Portuguese enclave, bearing arms with 'tanta graça' (as much grace) as the men. Nunez do Leão, now referring to Rodrigues in the masculine, noted that he became 'amado de todos os soldados' (loved by all the soldiers), in whose company he ate, slept and fought. Such a situation continued for over a year. Given his skills in horse riding, Rodrigues assumed the position of cavalier, defeated many Moors and became known as the best horse rider of the troops stationed at Mazagão. After five years in this role and in receipt of much admiration from his fellow soldiers and the female population of Mazagão, Rodrigues, fearing discovery, decided to confess to the priest of the locality, who informed the governor about his 'real' identity. As a result, he was forced to assume women's clothing. Once Rodrigues's femininity was confirmed, a marriage with an honourable knight ensued. As a reward for her services to the Crown, she was provided with a pension and a supply of grain by King Philip of Portugal.[8] On returning to Lisbon, now aged thirty-five, she came to the attention of the chronicler, Nunez do Leão, who described her as 'bem parecida' (good-looking), with 'muita graça' (much grace) and with 'grande viueza de spirito' (great liveliness of spirit). Such were the achievements of Rodrigues, Nunez do Leão suggested, that a statue should be erected in Mazagão to the glory of this 'female knight'.

What is arresting in the account by Nunez do Leão, in similarity to other stories circulating in the same period and in ensuing years in Spain and Spanish America, was the wholly positive judgement of Rodrigues.[9] On one occasion

Nunez do Leão refers to her as 'o fingido Antonio Rodriguez' (the supposed Antonio Rodriguez), but this is not a qualitative evaluation or the censoring of any attempted deception on the part of Rodrigues. Rather, it is in the context of the praise steeped upon her for her acts in defence of the realm. Furthermore, the account by Nunez do Leão shows how moveable sex roles were in the 1600s. As we have argued elsewhere, it is important to signal the assumption of status and the taking of a male role, apparently irrespective (until discovery or revelation) of femaleness, and it is this that makes the man. The appropriate dress[10] and actions, particularly in the military sphere but also in respect of physical work (Rodrigues's ability to run up the sails and climb the mast of the ship was noted), suppressed any doubt as to her identity as a woman. Once this was discovered, she changed back to inhabiting the status set out for women at the time. This reversal was not seen by Nunez do Leão as Rodrigues gaining once more her 'true' sex, and his account does not venture such interpretations.

The case of Rodrigues, in this respect, is different from cases of suspected hermaphroditism, which, after the appropriate medical examination, declared the subject to be either male or female; there were no such doubts in her case, and once she had revealed that she was a woman, she reverted back to the role and dress prescribed for women in the early 1600s. Questions of this nature and as to the 'true' sex of Rodrigues did, nevertheless, emerge during later stories, but the framework for such nineteenth- and twentieth-century accounts was fundamentally different from those operating more than two hundred years previously. In addition, it was only in subsequent versions of her life story that the dangers of possible same-sex encounters were highlighted. Although this was not an explicit element in the account given, for example, by the Count of Sabugosa, several references were made in his version to a young knight who found 'António' Rodrigues's looks captivating and who was clearly in love with the young soldier. Perhaps even more remarkable at the time was the lack of condemnation and even commendation by the king despite the existing laws on cross-dressing in force at the time (discussed below). Once again, the explanation for this can be found in the exceptionalism of Rodrigues's case: she dressed as a man to combat the Moors and, once advised to do so, reverted back to the dress 'appropriate' to her sex. Given this twin set of circumstances, any legal sanction was thus avoided. There was no association with sodomy either; despite contemporaneous cases brought by the Inquisition for the nefarious vice, Rodrigues's activities were of a different social and significatory register.[11] The ocean-going dimension of the tale is, of course, compelling, not least for what it may suggest about the acceptability of sex change among the population at the time, but also because of the prevailing background provided by the transformational stories of strange sea creatures and different types of humans encountered on such voyages.[12]

The story of Rodrigues, as the anatomist Pires de Lima noted, may well have been at the heart of the sixteenth-century romance of a girl who goes to war and is only identified as a man by her captain on his noting the vividness of her gaze. This romance, recorded by the nineteenth-century literary figure Almeida Garrett under the title 'The Girl Who Went to War', was perhaps originally a Castilian story also known in Portugal as 'The Boy of Count Daros'.[13] Once the disguise was discovered, the girl in question married her captain. Although Pires de Lima argued that this romance was based on a 'formosa lenda [que] tivesse real fundamento' (beautiful legend [which] had a real basis),[14] that is, the Portuguese cavalier who fought in Mazagão, what is certain is that such stories were recounted in the oral culture of the time and lasted into the twentieth century as cases of sexual ambiguity or, for Pires de Lima, a case worthy of mention in his analysis of hermaphroditism.[15]

Medieval romances, in which women sought to elude forced marriage, and early modern tales of seafaring and military heroism, where they dressed as men to achieve their aspirations for fortune or adventure, were common enough to catch the eye of chroniclers or to be recounted as romantic stories; but what was the legal framework on cross-dressing in the sixteenth century, and how did this inform discussions on hermaphroditism? The biblical prohibition against dressing in the garb of the other sex is well known, and it appears that around 1600 prohibitions in Iberia on such antics hardened. While cross-dressing in both Spanish and Portuguese theatre (as well as in other European theatres) was somewhat of a tradition, the Jesuit Juan de Mariana argued in his *Tratado contra los Juegos* (1609) that such a practice was dangerous and that the employment of female actors over boys who took female clothing was preferable.[16]

As we pointed out in Chapter 2, such an attempt to tighten the theatrical codes was made in the context of a more generalized concern over 'effeminacy' in the Habsburg period as the brilliant days of the Empire were beginning to appear less lustrous, inaugurating 'a vigorous discourse in seventeenth-century Spain that tried to restore a code of proper manhood'.[17] Within this new shift in discourse on manliness, it was not the focus on masculinity that was new but the association between weak masculinity and the experience of decline that was innovative.[18] Fray Juan de Santa María, the confessor of the Infanta María, the daughter of King Philip III, warned against the 'corrupcion de costumbres que los varones se regalan, y componen como mugeres' (corruption of customs whereby the men make themselves up and act like women).[19] Overindulgence in food, lack of sexual restraint, poor economic management and the wearing of high collars and long hair all became associated with effeminacy and even, occasionally, sodomy,[20] manifestations of womanliness in men that were not lost on the Portuguese, if early eighteenth-century commentaries are to be believed.[21] Such a range of concerns led to repeated legislation banning women from assuming male attire in Spain (six different laws and edicts between 1608 and 1675), and in Portugal the Ordenações Manuelinas, Ordenações Filipinas and Leis Extravagantes condemned the practice for both sexes from the sixteenth to the nineteenth century.[22] Unsurprisingly, the practice also came to the attention of the officers of the Inquisition, above all if it was accompanied by accusations of sodomy, diabolical pact or hermaphroditism. The black slave António, who took the female alias of 'Vitoria' and worked as a prostitute, was accused of sodomy and examined for possible hermaphroditism in 1557.[23] Despite Antonio stating that he was a woman, not a man ('sou negra e não negro' (I am a black woman, not a black man)), any evidence of female organs was rejected and he was condemned for sodomy.[24] If prostitution and disorder could entail the attention of local society and the Inquisition, so could 'transvestism', which served as a cipher for hermaphroditism. Felipe de la Torre was questioned by the Lisbon Inquisition in 1638 for his donning of female clothing and possible sodomy.[25] What was the worldview among theologians, doctors and officers of the Inquisition that drove such understandings and procedures? The next section sets out these debates by focusing on the thought of important Portuguese scientific figures such as Amatus Lusitanus, Rodrigo de Castro, Zacuto Lusitano and Isaac Cardoso with respect to their ideas on sex change and hermaphroditism. Their thought, in turn, is set against the evolving Inquisition in the peninsula and, in particular, the Portuguese variant, established in 1536.

Science, Sex Change and the Inquisition

The establishment of the Inquisition in Spain was followed closely by the order for the expulsion of the Jews or their mass conversion from 1492 onwards. The mass baptism of Jews in Portugal from 1497, as decreed by King Manoel, affected those Portuguese and Spanish Jews who had left Spain in 1492. The Portuguese Inquisition came in 1536, and this resulted in the exit from Portugal of numerous philosophers and scientists overseas.[26] Of the 50,000 Jews who left Spain after 1492, it is likely that the majority, from central and southern Spain, went initially to Portugal and North Africa.[27] Naples was a favoured destination, second to Portugal and then Amsterdam around the year 1600.[28] After the introduction of the Inquisition in Portugal and further subsequent anti-Jewish sentiment, many Portuguese-born Jews or converts in turn left for cities such as Venice, Verona, Antwerp and Amsterdam.[29] Fernando (Isaac) Cardoso was typical of this kind of migration. His parents were Spanish *conversos* who lived in Portugal until 1610 when they returned to Medina de Rioseco.[30] He went to Italy in 1648 and settled in Venice and then Verona, publishing his *Philosophia libera* in 1673 in homage to the Venetian authorities.[31]

The establishment of the Inquisition by papal bull in 1478 in order to root out remaining Jewish influence among the *conversos* took on an extended dimension from the late fifteenth century, when eyes were turned not only on internal threats to the Christian Church and the order of the early modern monarchical states that were being articulated, but also on other 'agents of Satan' in overseas territories. The significance for Spain of 1492 was not, therefore, only that of the end of the Reconquest and the 'discovery' of America but also Christianization in the new lands and the establishment of the Inquisition in the Iberian dominions.[32] In Portugal this move was also accompanied by the new valorization of science, especially astronomy, astrology and sea-faring knowledge, and an important development of medicine and plant science in Portugal and in the 'New World' which emphasized observation and empirical study within a predominantly Galenic and Hippocratic model.[33]

The circulation of this knowledge throughout Europe and its translation into different languages need no further elaboration.[34] The role of Europe's Jewish population in the positing of new understandings of nature has also not been lost on commentators.[35] As central European Jewish intellectuals in the sixteenth century argued that the realm of nature and the realm of the divine were distinct spheres, this opened up the way for different conceptualizations of the physical world, of which humanity formed a part.[36] Given this context and the search for new answers to old questions about the variability of nature, what were the understandings of Jewish or *converso* experts on hermaphroditism?

In the next section, the thought of a number of Portuguese-origin figures such as Amatus Lusitanus, Rodrigo de Castro, Zacuto Lusitano and Isaac Cardoso is traced with respect to their contribution to knowledge on sex change and hermaphroditism.

Amatus Lusitanus

João Rodrigues de Castello Branco (Amatus Lusitanus) constitutes a good example of the kind of work coming out of the opportunities presented in the post-expulsion period. Amatus Lusitanus was born in Portugal in 1511 of 'marrano' parents who had come from Spain. He studied in Salamanca, where he took his degree, probably in 1532. Conditions in Portugal drove him to Antwerp in 1533, and he took a chair in medicine in Ferrara, arriving there in 1540. In 1547 he took up residence in Ancona. After anti-Jewish repression, he fled and passed some time in Ragusa (now Dubrovnik) and then went to Salonica, where he died in 1568. His main work was a seven-volume collection of medical case histories, the *Curationum medicinalium* (published from 1551 onwards) with commentaries and discussions. Each volume contained one hundred cases (a *centuria*).[37]

Widely cited, for example by Ambroise Paré in his book on monsters and prodigies with reference to his thought on hermaphroditism,[38] Amatus also acted as a conduit for the reintroduction of Arabic thought on science by citing and admiring the work of his contemporary, an Italian Jewish physician, Jacob Mantino, who had been born in 1490 in Spain. Mantino had translated Averroes and Avicenna into Latin.[39] Although parts of the work by Amatus were placed on the Index by the Inquisition, resulting in the censorship of some sections particularly related to questions of sexuality,[40] his comments in *Centuria II* on the question of hermaphroditism and the possibility of sex change in human beings can be observed in editions preceding such interventions.[41] While the Burgos edition of 1620 of this work was censored precisely in Cure XXXIX on the subject of hermaphroditism, an earlier edition is intact.[42]

Amatus's understanding of sex was characteristically Galenic and Hippocratic, authors who are cited abundantly in the work, and 17 per cent of cases studied which made up questions related to sexuality in the *Curationum medicinalium* dealt with issues such as the physiology of the reproductive parts, the foetus, disorders of the womb, conception, semen, etc.[43] As we have pointed out in the case of some Spanish writers during the same period, however, although Amatus accepted the Galenic understandings of semen as produced by both men and women and, primarily, of the reproductive organs of the sexes as being largely the same (with the male organs being understood as placed outside the body and the female apparatus placed within the body), he also conceded some small but important anatomical differences between the two sexes. These went beyond the differences signalled by Galen in respect of the function of male and female semen and the differences in temperature between men and women. Amatus did not highlight the inferiority of female semen as merely a means of providing feed for the foetus, although, like Galen, he admitted that it was the blood that produced both male and female types of seed. When discussing the vessels channelling the semen to the pudenda, Amatus noted that in the male there were two and in the female four tubes.[44]

Despite admitting small anatomical and organizational differences such as these, and despite questioning Hippocratic notions of the outward signs of sex determination by examining the pregnant mother,[45] the ideas of Amatus on sex difference largely coincided with those of his forebears. In this light, and in contrast to Aristotelian thought, Amatus accepted that hermaphroditism and sex change were possible within a model that supposed the movement from femaleness to maleness as an 'improvement' and advancement in the human type. Being contemporaneous with Ambroise Paré, Amatus made numerous references to ambiguous genitalia, sex change and hermaphroditism.[46]

In the first *Centuria*, Amatus refers to the case of a two-year-old child whose urinary opening was situated near the testicles rather than in the penis. This reality led the author to declare that the child's nature showed participation in 'naturam & masculum & fœminam', suggesting 'ex hermaphroditorum genere'.[47] A surgical remedy was proposed by doctors, but the parents would not consent to such 'differentiation'. This meant that the child would perhaps have lived between the sexes at least in its early life.[48] Male haemorrhoidal menstruation was also alluded to by Amatus in the case of one subject, João Baptista.[49] The causes of such 'menstruation' in men were, however, more often put down by him to the consequences of anal sex rather than a specific quality of men.

The most significant case discussed by Amatus of possible hermaphroditism was the one recounted by Paré: that of Maria Pacheca. As well as being recorded by Paré and other Portuguese physicians of the period,[50] the case was referred to in Portuguese medical circles well into the twentieth century.[51] The case of Maria Pacheca from Esgueira, which came to light during Amatus's life, was referred to in his *Curationum medicinalium* as an example of sex change. Pacheca, on arriving at puberty, developed a penis that had hitherto been hidden. In this way, she changed sex and took the clothes of a man and adopted the name 'Manuel'. Pacheca then went to India, made his fortune and returned to begin a family. Throughout this process, Amatus noted, he remained beardless. Despite this lack of an outward sign of masculinity, in the explanation of the case Amatus accepted that it was no myth that women could be transformed into men ('Ex fœminis mutari in mares non est fabulosum') and drew on Pliny and Hippocrates to buttress his claim.[52]

Roderici a Castro Lusitani and Stephani Roderici Castrensis Lusitani

Rodrigo de Castro (Roderici a Castro Lusitani) was born in Lisbon in 1546, studied at Salamanca and then practised medicine in Lisbon. He subsequently left Portugal and settled in Hanover. Among his publications was *De universa muliebrium morborum medicina* (1617), an influential gynaecological text of the seventeenth century which discussed the possibility of hermaphroditism in light of the thought of Galen, Aristotle, Hippocrates and Avicenna.[53] Four main types of androgyne or hermaphrodite were discussed along the male/female scale, and the legislation with respect to hermaphrodite individuals was noted. His work is not to be confused with that of Stephani Roderici Castrensis Lusitani (Estêvão Rodrigo de Castro, Lisbon 1559–1627), who also in his *Tractatus de natura muliebri* elaborated upon the possibility of sex change (considered below).[54]

In Rodrigo de Castro's four-page discussion of hermaphrodites in *De universa muliebrium morborum medicina*, in the same way as a minority of contemporary commentators such as Luis Mercado and unlike Cardoso, he confirmed the categorization of hermaphrodites as a species of monster, of which the hermaphrodite was the commonest kind.[55] Despite ample discussion in the work of Hippocrates and Galen, Castro argued that the lack of attention in gynaecological treatises justified his current intervention. He carefully distinguished between different versions of ancient Greek thought on the subject, noting that while Aristotle believed that two sexes could occur in the same individual, only one of these was 'valid' for this author and the other, through lack of alimentation, was less developed, becoming a kind of 'abscess'.[56] As in Spain at the time, the presence of Aristotelian thought coincided with Galenic-Hippocratic models (see Chapter 1), although the one-sex model broadly dominated. Among the four kinds of androgynes outlined by Castro, a sophisticated repertoire was constructed, not commonly seen during the period. In this, however, he did coincide broadly with the four types of hermaphrodite identified by his fellow countryman Pedro de Peramato in his *Opera Medicinalia* (1576).[57]

The first type of hermaphrodite referred to by Castro was a male who possessed a full virile sex but who had a perforation in the perineum. No semen or urine, however, came from this place, and this type was often confused with those suffering from hypospadias. The second full female type possessed above the vulva an excrescence that was similar to the penis. This type was described as a 'nymph', and the subject did not have any testicles and could not produce semen. The third type possessed the characteristics of the two sexes but was incapable of producing semen of either type, although urination was possible. The fourth type in Castro's repertoire included those that could use the organs of both sexes with equal 'potency'.[58] Castro instantly qualified this categorization by drawing on other interpretations, including that of Aristotle and the seventh-century Paul of Aegina, to note that in this last type that might be termed the 'perfect hermaphrodite' (not a term Castro used), in reality, following Aristotle, one set of genitalia was strong and the other useless. Galenic thought was, on the one hand, employed to buttress claims for hermaphroditism; on the other, Aristotle was cited in order to disavow such a possibility in reality. Pragmatically, Castro argued that whichever sex predominated would designate the individual; those who possessed equal quantities of both would be a 'vir fœmina', and those who possessed insufficient of either, a neutral class.[59] Various causes of hermaphroditism were then outlined, including the theory of the seven-part womb, the influence of the day of conception and the Hippocratic notion of the relative strength of male and female semen.[60] Although Castro coincided with the fourfold classification of Peramato, who is not mentioned in his account, there are also differences. Peramato believed that hermaphrodites did indeed exist and that they were the origin of stories of changed sex rather than any other internal or external process of sexual transformation. Both Pliny and Hippocrates were referred to for this explanation of 'hidden hermaphrodites' and sex changes, such as the case mentioned by Amato from Esgueira, which was an example of this hidden nature predominating. In this sense, there was no real change of sex, according to Peramato ('non est permutatio sexu, ut scripsit Plinius'), but the work of the 'second sex' of the hermaphrodite ('sed occulti sexus detectio') came to the fore.

Estêvão Rodrigo de Castro, in his *Tractatus de natura muliebri*, declared from the outset that only sex change in one direction, from female to male, would be considered as the reverse had been dismissed as impossible. The 'perfection argument' was advanced once again as the explanatory frame for such an assertion. That men could become eunuchs, who were women 'in a certain way', was not denied, however.[61] Having dismissed the possibility and having accounted for those cases in which men dress as women in order to marry and commit 'nefandi concubitus', Rodrigo de Castro opened up a discussion on whether Galen found that female–male transformation was possible. Given the fact that Galen believed that the organs of both sexes were the same, a change in sex from male to female was not impossible in his thought. The imperfect organs of the woman would always seek greater perfection, and conversion into a man was thus feasible.

Despite Galen's (and Hippocrates's) acceptance of sex change, Rodrigo de Castro goes on to reject such a possibility. This was only feasible because of the author's explicit rejection of the one-sex model: anatomists had proven that there was no complete similarity or identical nature of the male and female parts, he averred.[62] This led expert anatomists to declare that both male–female *and* female–male transformation was impossible: these were 'chymerical conversions'. Despite Avicenna, Albucasis and Andreas Laurentius (André du Laurens) referring to a 'female penis' or the similarities between the male and female parts, a man's penis cannot be made from this clitoris. Rodrigo de Castro quite clearly endorsed what he understood as fundamental anatomical differences between men and women in this sense. Furthermore, the scrotum and the womb, rather than being made of the same material as Galen would have it, were made of 'completely different' matter.[63] As Cardoso suggested (see below), woman was made by nature to perform particular acts (reproduction), and nature could be altered in order to change sex.

Zacuto Lusitano

Zacuto Lusitano (b. Lisbon, 1575–1642) was another Portuguese-origin doctor who, by contrast, did accept the possibility of transformation from woman to man. In the sixteenth question of his *De medicorum principum historia*, he broached the issue of hermaphroditism in accordance with the thought of Galen, Bauhin and others.[64] The question was discussed as a philosophical debate whereby Pliny's understanding that such a transformation was possible was contrasted with the thought of Luis de Mercado, Laurentius and Severinus Pinaeus. Zacuto admonished these authors for their error in assuming that such cases were in fact examples of women whose clitoris was so large it was confused with the male penis. The reality was that these were often, in fact, cases of sex change. That such a phenomenon could exist was hardly surprising: other 'more monstrous' transformations were provided by Nature such as new teeth in octogenarians and the opening of the nasal and auricular orifices in babies born in the seventh month, as Avicenna, following Aristotle, had argued. What then, Zacuto asked, impeded the production of a virile member in a woman when she was being formed and which, at a particular moment, became manifest once fully developed? Given the propensity of 'errors' in Nature, how could it be suggested that a woman could not become a man? Furthermore, as Pliny, Aristotle and Avicenna had illustrated, hermaphrodites, as dual-sexed individuals, had been amply documented. It was also possible for women to grow facial and body hair, a factor that also converted them into men. Hippocrates, Alexander Benedictus and Avicenna, once more, were cited to argue this point as well as the case of Faetusa, a servant mentioned by Hippocrates, whose 'body was transformed through this cause into that of a man and she became hirsute all over'. If excessive heat made women grow bodily hair, why was it to be denied that 'the genitalia hidden in the interior of the body of women, on receiving the influence of heat outwards, may not come out to the exterior in the course of aging?' Numerous commentators in the peninsula had remarked on the same phenomenon, Zacuto noted: Torquemada, Martín del Río and Amatus Lusitanus had agreed such a possibility.[65]

Zacuto, in this way, subscribed effectively to the 'one-sex' model which viewed the genitalia in a developmental way and which allowed for sex change as part of a natural process within a schema that admitted the 'errors' of Nature. He also, as is to be expected, argued that it was impossible for a man to turn into a woman, as it was rare that there was a passage from perfect to imperfect. Various authorities were mentioned should the reader wish to examine this further.[66]

Isaac Cardoso

Among the gallery of Portuguese/Spanish philosophers and physicians who wrote of these issues, mention should be made of Isaac Cardoso (b. Trancoso, Beira Alta, 1603/4–80), physician to Philip IV, whose *Philosophia libera* also addressed the issue. As Yerushalmi notes,[67] Question 14 of this work attends to the question of sex change under the title 'De Sexus mutatione'.[68] Further, hermaphrodites are specifically discussed in Question 15, 'De Hermaphroditis'.[69] Cardoso, writing slightly later than Zacuto Lusitano, displayed some developments in respect of adherence to the one-sex model and the origin and status of differences between men and women. The understandings adopted in *Philosophia libera* drew extensively on European sources and particularly Spanish ones, relying on Torreblanca and Del Río as well as an eclectic mix from Hippocrates and Galen. Like Amatus Lusitanus, he noted that even though Galen had argued that the organs of both women and men were the same in structure but differed in terms of heat and position, for example, the actual nature of certain sets of apparatus was in fact different. In order to trace his argument, we focus first on his discussion of sex change and then on his shorter intervention on hermaphroditism.

Drawing on a wide range of ancient Greek stories and philosophical discussions, he argued that the impression had been given in the past that women could change into men. The reverse was not contemplated by Cardoso. Hippocrates, Titio Livio, Pliny and Ovid were marshalled to prove this point.[70] Cases from Rome and Naples, derived from Marcelo Donato and Fuigosius, respectively, served to introduce the case of Maria Pacheca from Esgueira as related by Amatus Lusitanus.[71] Other cases discussed in previous chapters were mentioned by Cardoso, such as Magdalena Muñoz from Úbeda and the 'lieutenant woman', no doubt Elena de Céspedes, the latter having been known to the Portuguese author from his time in Madrid.[72] Cardoso had also observed a 'semi-bearded' Italian who had been a woman before, had given birth to two sons, dressed delicately and possessed 'soft flesh', at the palace of the Prince of Esquilache, Francisco de Borja y Aragón. Other examples of intersexuality peppered the text by Cardoso. He discussed how Brígida de Peñaranda was described by Zacuto and how cases of bearded women were analysed in the work by Martín del Río, *Disquisiciones mágicas*. All these textual and observational examples allowed Cardoso to conclude that there was no doubt that women could change into men, but precisely how this was possible was 'highly uncertain' and it constituted a design that was hidden in the mysterious folds of Nature. But Cardoso did not agree entirely with the theories of Galen on complete sameness between men and women apart from the position of the bodily organs. In addition to the difference in respect of position, the organs of the two sexes also differed in terms of their 'conformation'. Women's genitalia were not hidden simply because they were colder; rather, they were hidden because it was in their nature to be so. Other differences were observed: the seminal tubes were in a different place (this difference seemed to be merely positional), but women's bones were broader and wider in order to sustain the foetus. Any externalization of the male genitalia in cases of sex change was not simply due to the excessive heat contained in them but because of their very nature pushing them to the outside. Cardoso was cautious, nevertheless, like many of his contemporaries as we have seen in previous chapters, to distinguish between cases of sex change and cases of women whose large clitorises caused them to be confused with men. This in turn could be due to some kind of illness or distortion, and even though such a large clitoris could be employed as if it were a penis, there was no seminal discharge. Perhaps, Cardoso admitted, these women were more androgynous from the outset, with one sex visible and the other hidden, and as nature took a different turn the masculine element popped out.[73] A large range of 'in-between' animals and birds and those with hidden or dual genitalia was cited to prove such a case in nature, and both Huarte de San Juan and Torreblanca were drawn upon to back up such claims.

Despite Cardoso's acceptance of some differences between men and women that went beyond mere temperature and the position of organs, and in contrast to his overt acceptance of the possibility of sex change from woman to man, the reverse was explicitly rejected as a possibility, either in nature or by art of the devil.[74] This was not only because nature advanced from the imperfect to the perfect, but also because the genitalia would only emerge because of heat and never because of the effects of cold. This somewhat circular argument was still, evidently, rooted in the differential qualities of warmth and coldness that Cardoso himself had drawn upon previously. Any cases of supposed conversion from men to women were in fact due to confusion or lack of observation; any such change was more apparent than actual.

This ameliorative framework, proceeding from woman to man, that acknowledged differences in bodily temperature and also differences in the position and nature of the organs that made up men and women, was of significance to Cardoso's next subject: hermaphroditism. As in his discussion of sex change, Cardoso introduced his analysis by means of examples taken from mythology, religion and philosophy. The hermaphrodite, although acknowledged to possess portentous charm in the past, was differentiated from the monster in human and divine law, and Aristotle was criticized for his 'lax' use of the term.[75] Four types of hermaphrodite, three of which would be male, were then identified. Within this classification, there may be a certain male or female predominance, there may be an equal balance, or there may be no predominance at all, and the majority are conformed by female genitalia which exist below the male apparatus. Following Aristotle, Cardoso argued that there is always one useful and one useless sex in any such individual and in the majority of cases neither sex is 'useful', although there may be predominance in some cases. He pointed to the work of Zacchia, *Quaestiones medico-legales*, to argue that there have been no cases of hermaphrodites that have been both fathers and mothers or in whom there is a perfect harmony of maleness and femaleness. Following Avicenna, the semen cannot possess two 'contrary movements'.[76] From this analysis, Cardoso concluded that although not strictly impossible, it is highly unlikely to find a hermaphrodite that can generate from both sexes. Numerous examples from the animal world follow, and a final exposition on Jewish law, drawing on Bauhin, states that male androgynes can take a wife as long as they dress appropriately; the reverse is not permitted. The latter can shave; such an act is not permitted for male hermaphrodites. What were the causes of such a phenomenon? As a coda to this section, after an account detailing various theories from natural history and ancient Greek explanations, the story of Adam and Eve is advanced to explain how woman comes from man and will return to man as part of the indivisible nexus between the sexes and the natural order.

While it was only Amatus Lusitanus and Cardoso who specifically recorded cases of hermaphroditism in Portugal, the influence of these physicians in respect of the dissemination and discussion of knowledge on sex mutation was important in the sixteenth and seventeenth centuries, and there was some mutual citation of their works.[77] In addition to these figures, and as an example of how widely it was understood that hermaphroditism or sex change was indeed possible, we can cite the work of Padre António Vieira, the Jesuit who not only defended the Jews against accusations of heresy but also raised his voice against the practices of the Inquisition. A defender of indigenous peoples in the 'New World', Vieira died in Brazil in 1697. He had also been the confessor of Queen Christina of Sweden (1626–89), who was accused of being a hermaphrodite, and it was Rodrigo de Castro who was her one-time physician. More concretely in respect of hermaphroditism, Vieira had written about an episode that had occurred in India where Saint Francis Xavier performed the miracle of transforming a recently born girl into a boy in response to the request and prayers of a father.[78] The possibility of such an event taking place was discussed in a work attributed to Vieira, the *Arte de Furtar*. In the sixteenth chapter of this book, he acknowledged, when discussing inheritance and family relations, that Nature had made many women into men, something that was easier than making a son possess the age of his father.[79]

Gaspar dos Reis Franco

Finally, worthy of mention is the extensive volume *Elysius iucundarum quaestionum campus*, written by Gaspar dos Reis Franco (Gaspare a Reies Franco) in 1661.[80] The importance of Reis Franco with respect to the question of hermaphroditism was highlighted by Pires de Lima in his 1939 overview as an author who believed, alongside Zacuto Lusitano, in the possibility of 'mutação dos sexos' (changes in sex).[81] Reis Franco was educated in philosophy at the University of Évora and as part of his exploration of philosophical and medical matters addressed the issue of desire and potency in androgynes and hermaphrodites.[82] Under Question 48 in his treatise, entitled 'Can Androgynes and Hermaphrodites Who Can Use Both Sexes be Found?', the author engages in a long discussion on the existence or otherwise of hermaphrodites from ancient times to his period of writing. Like his contemporaries in Portugal or those of Portuguese origin, the erudition of his work is striking, and after a survey of Greek mythology and the works of Pliny, Paul of Aegina, Cicero, Aristotle and Avicenna, other more unusual authors are cited such as Rhodiginus, Schenck and Aldrovando to assert that hermaphroditic animals and humans had been found, and an examination of the causes of such phenomena followed.

The very possibility of hermaphroditism was based on Reis Franco's admission that some doubt existed about whether men could lactate: would these be women or men who somehow participated in the female sex? Such a conundrum invited an exploration of whether hermaphrodites could use both sexes, that is, whether they 'were capable of taking the active and passive role, could inseminate other women and, in addition, whether other males could inseminate them'.[83] A variety of causes of hermaphroditism was then discussed, from what Reis Franco dismissed as the erroneous interpretation of the supposed biblical androgyny of Adam (an expression of heresy he identified as developing during the time of Pope Innocent III),[84] through to the medical explanation of Suessanus, who argued that up to the fifth day after coitus a male would result, up to the eighth a female, from then until the twelfth a male, and from this day hermaphrodites would be born.[85] This interpretation was rejected by Reis Franco, who drew on Castro's *Tractatus de natura muliebri* and Aldrovando's work on the history of monsters to assert that they correctly, in his view, denied such a possibility. A range of other traditional explanations was also offered such as the position in the womb where conception takes place and the relative power of the semen in the formation of the new individual.

The notion that the imagination alone could produce a change in sex in the womb or a hermaphrodite was also questioned. While, Reis Franco argued, this could be a factor in two types of hermaphrodite, in the fourth variety where the genitals of both sexes were perfectly formed and were capable of engendering, this explanation was less viable.[86] The addition of a sex or sex change could therefore be explained by the later arrival of male semen, which would result in the presence of two sexes, as argued by Joannes Benedictus Sinibaldus. A whole set of authors was cited by Reis Franco to back this claim.[87] In such a scenario, neither of the two types of semen would be able to impose itself on the other as both possessed similar levels of energy. Hippocrates had argued along these lines and had also explained the existence of men of 'exceptional masculinity' and those who were effeminate or androgynous depending on the relative strength of the two types of semen and the battle that ensued to impose one on the other. Aristotle was also interpreted in order to sustain this explanation.[88] The varieties of hermaphrodite were also to derive from such a scenario, although the fourth type, the fully functioning hermaphrodite, was extremely rare.

For Reis Franco, this variety, it is important to note, not only existed but was also capable of the complete use of his/her genitalia and could reproduce.[89] This circumstance, however, was denied by Aristotle and by Marcelo Donato. Although cases had been seen of women with large clitorises, reported in work by Zacutus among others, confusing the issue, Reis Franco was still keen to assert the possibility that this multiplicity and complexity could exist within one individual with full capacity. Nature in all her majesty was well disposed towards 'games' or tricks of this type.[90] The results could be seen in many human societies in Africa and India and in various parts of Europe; one Spanish case in Burgos discussed by Torquemada and another in Seville were evidence of this.[91]

It was noted, however, that Aristotle was not of this opinion. Reis Franco argued that any opposition or exclusivity between the sexes was in fact not a matter of 'incompatible opposites' but an opposition that existed between the semen of one and the semen of the other sex.[92] Nature produced monsters on a daily basis; what was there, therefore, to prevent the creation of hermaphrodites? Having established the possibility and some of the causes of hermaphroditism in an extensive analysis of a variety of sources, Reis Franco asked what legal framework should govern their existence. The very fact that the laws governing which sex should be used existed was proof of the juxtaposition of two sexes in one individual, as Avicenna, Castro, Zacuto Lusitano and Zacchia had argued.

In addition to displaying how the thought of Aristotle was certainly not hegemonic in the seventeenth century but was considered alongside multiple other explanations, Reis Franco argued that it was necessary to examine the hermaphrodite in order to concede a particular legal statute of sex for the individual concerned. This would depend on the 'conformation' of the genitalia and the ability of the individual to exercise the appropriate role. Legal experts, bishops and doctors would prescribe the sex, and it would be necessary for the hermaphrodite to elect one sex, the sex in which s/he felt most comfortable. Once this sex was chosen, the individual would swear to keep to it; infractions would result in severe punishment. Such had been established by a number of legal experts, including the Spaniard Sánchez in his volume on marriage (see Chapter 1). The work of Reis Franco is interesting in this sense and would appear to allow a degree of agency for the hermaphrodite in choosing his/her sex. Furthermore, it was possible, Reis Franco noted at the end of his long analysis, that the individual could become more 'attracted' to the other sex (that is, more comfortable with this sex; we are not talking about 'homosexuality' here) and that 'the heat of his/her nature could change'. A recipe for the resolution of such a scenario was not advanced, although he noted that Majolus in his *Colloquium* had dealt with such a possibility.[93]

Cases of Hermaphroditism in Portugal

While some of the cases of hermaphroditism that have been discussed above were dealt with by the Holy Office or the ecclesiastical authorities, not all were, or at least were not initially. The files of the Inquisition, in addition to the incidences of hermaphroditism that we have already detailed, nevertheless, are evidently a primary source in this sense.[94] In this last part of the chapter we will consider the following cases of suspected hermaphroditism, often with the accompanying accusations of sodomy, which were considered by the ecclesiastical authorities under one guise or another: those of Sister Claudiana da Natividade (Vila Viçosa, 1622); Manuel João (1637);[95] Estêvão Luís, 'o Cobra' (the Cobra) (1686);[96] Father Pedro Furtado (1698–1701);[97] and finally that of José Martins (1725), the 'she-man' of Ervedal.[98]

As we have seen, the triangle of associations between heresy, sodomy and other forms of sexual transgression provided a potent focus for considerations on hermaphroditism in the period we are discussing.[99] Accusations of sodomy were numerous during the period of operation of the Inquisition in Portugal, which was based in three principal localities: Lisbon, Évora and Coimbra. As Luiz Mott has identified, there were some 4,419 persons accused of sodomy, of which 394 were held prisoner and charged over the period 1587 to 1794, representing just less than 9 per cent of the total cases heard;[100] also heard were some cases of lesbian sodomy. Of most relevance for this study is the overlap in charges for sodomy with, for example, witchcraft, sorcery, deceit and the accusation of pacts with the devil. It is in these cases that the question of hermaphroditism or sex change often appeared, that is, in circumstances where there was the accusation of deception or unnatural activity, whether through a supernatural pact, counter-natural sexual activity (sodomy) or the attempt to present oneself as something that one was not, thus breaking with the condition of rank, sex or 'gender' operating at the time.

The first case we will consider was rescued from oblivion by the medico-legal expert António Asdrúbal de Aguiar and was referred to at the start of the Introduction of this book.[101] De Aguiar detailed the accusation of hermaphroditism dating from 1622 against Sister Claudiana da Natividade, a nun at the Santa Cruz Monastery at Vila Viçosa, in a five-page account published in the mid-1920s.[102] Before discussing the case of Sister Claudiana as a 'case of female pseudo-hermaphroditism' in detail, de Aguiar provided an overview of scientific and religious thought on the subject from the sixteenth and seventeenth centuries. Far from resolving a series of questions about hermaphrodites, from the legal aspects of such cases through to the 'dangers' such individuals may have posed to others, these inquiries provided nothing definitive, he argued; from them 'brotavam novas dúvidas, nasciam novos embaraços e enredava-se mais o assunto' (emerged new doubts, new conundrums were born and the issue became more and more complicated).[103] Attention was paid as to whether hermaphrodites could inherit, whether they could marry, whether they could practise medicine; particularly relevant to the Portuguese case in hand, de Aguiar noted, was whether the individual could become a nun or priest and whether they could live in a convent.

The initial source of consternation in the Santa Cruz monastery was the 'comportamento esquisito' (strange behaviour) of the nun, which had gravely undermined proper custom in the religious setting: Claudiana was attracted to the other nuns, although it was not known whether any of the nuns or novices had been the object of any 'especial atracção' (special attraction) from her.[104] The mother superior, uncertain as to what to do, reported the case to Friar Jorge de Sande, 'provincial' of the Saint Augustinian Order of Hermits in Portugal. Friar Sande ordered some matrons to examine Claudiana and organized the gathering of testimonies. The matrons pronounced that the body of Claudiana 'mais parecia na sua natureza homem que mulher' (appeared in its nature more like that of a man than that of a woman), and witnesses declared that she alleged that she was the victim and not the perpetrator of any scandalous behaviour, although she did not deny her 'abnormality'. On hearing her confession, Friar Sande decreed that Claudiana should be removed from the Order. The sentence against Claudiana, 'por ser homen, e não mulher' (for being a man and not a woman),[105] was then carried out in 1622.

In his declaration, Friar Sande outlined the motivations that had brought Sister Claudiana to the convent in the first place. Although always problematic in respect of the reliability of such a second-hand account, there is no particular reason to doubt the veracity of Claudiana's admission that she became a nun on becoming aware that she 'não tinha vazo natural como as mais mulheres para poder casar' (did not possess a natural vessel like that of other women in order to be able to marry).[106] In addition to this act of self-reflection and proposed remedy for her 'abnormality', she admitted that she had been examined visually and physically on numerous occasions and was found to possess a man's nature. This she sometimes admitted to others; at other times it was a reality that she kept hidden. Through confession she told her priest that she underwent 'deleitosas seminis effusivas' (pleasurable effusions of semen) as would a man.[107] In addition, this man's nature was sometimes retracted into the body but remained outside when it became stimulated. It had been proven by examination that Claudiana had no woman's nature and no appropriate vessel to receive semen.

In a further indication of Claudiana's self-perception, rare in documents of this nature, Friar Sande recorded that on actually believing herself to be a man, she had asked for an appropriate remedy for her situation for the sake of her soul and her honour. Not only did Claudiana possess male genitalia; she also detested 'feminine' tasks and always tried to undertake what were considered to be men's duties in the convent. The sentence, of 16 December 1622, pronounced that Claudiana should depart from the convent within the space of two hours.[108]

De Aguiar remarks that if Claudiana had not confessed immediately, Friar Sande would have had no option other than to deliver her into the hands of the Inquisition. Although it is not known whether Friar Sande knew of other cases similar to that of Claudiana, such as that of Marin Le Marcis, who ten years previously in Rouen had undergone a similar process and had been condemned to death (later commuted after Duval's intervention, as de Aguiar points out),[109] what is clear is that Claudiana was more fortunate than the other cases we consider below who were handed to the officers of the Inquisition. However, Friar Sande's sentence did not spell the end of the matter. Claudiana appealed to the pope some fifteen years after her removal from the convent, Friar Sande's decision was revoked, and she was incorporated into the Order once again. No further incidents were recorded, and the nun ended her days dedicated to the 'prácticas místicas' (mystical practices) to which she was devoted.[110]

The same year that the pope's dispensation for Claudiana came through, in 1637 the cook Manuel João, also known as 'o Bicho' (the Animal), was accused of sodomy and brought before the Coimbra Inquisition. As François Soyer has pointed out, Manuel João worked in the seminary at Viseu and was arrested for sodomy by the inquisitorial tribunal of Coimbra in 1637.[111] As well as performing women's tasks, he spun thread at the spinning wheel (commonly associated with witchcraft). In his case, it was the 'transgression of the division of labour in the kitchen along gender lines that raised eye-brows, and eventually, suspicions, about his sexuality'.[112]

Estêvão Luís, 'o Cobra' (the Cobra), was a seventy-eight-year-old African freedman prosecuted by the Évora tribunal between 1686 and 1690 for suspected witchcraft and a pact with the devil. He was also accused of sodomy. As James Sweet has recorded, when Estêvão Luís was detained to answer the charges of witchcraft (for being a 'feiticeiro' or witch), another detainee accused him of sodomy.[113] The claims of sodomy, in Estêvão Luís's case of being the 'passive' partner, went back some thirty years. Some witnesses claimed that Estêvão Luís was not only 'feminine' but that he may be in fact a hermaphrodite. In part, as we will see with the next case discussed below, that of Father Furtado, this may have been an attempt by witnesses to distance themselves from charges of sodomy; if sexual intercourse had taken place in the appropriate vessel, the accusation would not stand. Numerous statements during the trial were made to this effect: Estêvão Luís was thought to urinate like a woman;[114] to possess a 'buraco' (hole) 'por diante' (in front);[115] to be a 'macho e femea' (man and woman);[116] to possess a 'vazo ... como mulher' (a vessel ... like a woman);[117] and to be a 'hermaphro-dito' in league with the devil.[118] The Inquisition ordered, as was common in such cases, an examination; but instead of finding any evidence of female parts, surgeons declared Estêvão Luís to have an apparatus no different from other men, that he possessed a male member and had never possessed female parts, and that he only had a 'prepostero', an opening behind and none in front.[119] As Sweet points out, this did not dispel the conception among some of his countrymen that he was indeed a woman. In some respects, he was, of course, given his ability to provide remedies for diseases: the 'cultural feminization rendered him more likely to be penetrated by the spirits. And he was therefore seen as a strong and important force in effecting change in the African and African-Portuguese community of Evora' as a 'feiticeiro'.[120] While most of the trial, which extends to over six hundred pages, was devoted to this aspect of Estêvão Luís's activity and the determination of whether sodomy had been committed, the question of a possible pact with the devil and hermaphroditism featured strongly in the witness statements and the declarations of the officers. But it was the sodomy and witchcraft charges that drove the related accusation of hermaphroditism and the examination of the accused's body by surgeons. Once sodomy rather than hermaphroditism was 'proven', as a punishment Estêvão Luís was flogged in the streets of Évora and banished to Brazil.

Many of the same associations – between sodomy, supernatural confabulation and the attempt to pass as the other sex – were present in the three-hundred-page trial of Father Pedro Furtado (1698–1701), a case heard by the Coimbra tribunal of the Inquisition. To this set of misdemeanours was added the question of whether Father Furtado was fit to hold office as a priest and to administer the sacraments. If he was a woman in reality, such was not permitted under ecclesiastical law. If he was a hermaphrodite, as we have seen in previous chapters, the question was vexed but he would not necessarily be impeded from doing so. As François Soyer has extensively recounted,[121] Father Furtado came to the attention of the Holy Office because of the denunciation of one António Simões, who claimed that he and the priest had slept together repeatedly and that the priest had stated that he possessed female genitalia instead of a male apparatus. Simões recounted another doubtful characteristic of the priest: 'he habitually mixed pulverized stone from an altarpiece with ground tobacco' for its libidinous effect.[122] There followed a string of similar denunciations from men from the locality. Another partner, Marcos Villares, claimed that Furtado had told him he was indeed a woman and that the pope could, under certain circumstances, ordain a woman.

Clearly troubled by this experience, the men involved were concerned on several levels. There was the concern about the possible commitment of an act of sodomy, although the men declared that they had inserted their members into Father Furtado as if he were a woman. They were concerned about his ability to act as a spiritual father in the local community; and, further, they were troubled about his use of artificial, perhaps supernatural, potions to enhance sexual desire. There were also allegations that he may have used a 'female instrument' in his sexual practices. Father Furtado himself had declared to many men that he was in fact a woman, had flattened his breasts and had grown a beard. He also claimed that he had borne a child who had perished and been buried in the churchyard. No evidence of this child was found.

The Inquisition was faced, therefore, with a cluster of serious allegations. The officers set about their task of gathering evidence and interviewing witnesses. Some of these were deemed unreliable in respect of the allegation that Furtado was really a woman, and the initial finding of the inquisitorial officers recorded that, despite having the reputation of being a woman, 'havendo nelhe as mays demonstraçoens naturaes do sexo masculino como são as barbas, secura dos peitos, e aynda estatura; e disposição varonil' (there being in him the clearest natural demonstrations of the male sex such as a beard, a flat chest, his stature and his masculine disposition),[123] he was most likely a man. As was common in this type of case, the Inquisition ordered a physical examination of Furtado in April 1698 to be conducted by a medical doctor and a surgeon and presided over by two inquisitorial notaries. The result of the examination was clear: Furtado was a man. The doctor in charge observed that the accused possessed 'o instrumento com que a natureza distingue o sexo' (the instrument with which Nature distinguishes sex). Furthermore, in order to dispel any other possible interpretations, the doctor observed that he 'tinha membro viril de homem sem sinal, ou demonstração alguãs de que se pudesse conhecer ser molher, ou Hermafrodita' (possessed the virile member of a man and that there was no evidence or any manifestation that he could possibly be a woman or a Hermaphrodite). The existence of testicles further proved his masculinity and his lack of participation as a 'sexo misto' (mixed-sex person).[124] The anus was also in the correct place, and there could be no confusion with a vagina.

Despite this somewhat conclusive account, in light of further witness declarations that stated that no evidence of a male apparatus had been encountered in their relations with Furtado, the officers of the Inquisition began to doubt the veracity of the physical examination conducted by their own appointees.[125] This resulted in the Inquisition overturning the findings of the examination on the basis of witness declarations and rumour. Such a step not only shows the power of popular belief in the female nature of Furtado but also attests to the flexibility that many would give to possible movement between the sexes, despite some outward signs of masculinity, such as the beard and, in this case, Furtado's occupation as a priest. It should be noted, however, that these outward signs were only susceptible to inverisimilitude once a sufficient body of evidence had been accrued: it was precisely the stability of rank (in this case, as a priest) or its unmasking as false that provided for this possibility. To this was added a further factor: the possibility of Furtado having changed sex as a result of demonic intervention.

This was, of course, a serious charge for any man and was especially so for a man of the cloth, and it compounded the allegations that he had acted contrary to the precepts of his office. For some time, the inquisitors argued, he had 'esquecido de sua obrigação se fingio mulher por operação diabolica prouocando pessoas de sexo masculino a que tivessem com elhe Copula, como homen com mulher, e usava de outras superstições em grande deprejuizo de sua alma, e escandalo dos fieis' (forgotten his duties, had pretended to be a woman by means of a diabolical pact and had made persons of the male sex copulate with him, as a man with a woman, and had engaged in other superstitious acts to the prejudice of his soul and much scandal of the faithful).[126] This provoked a second physical examination in order to ascertain the sex of Furtado and to see if there were any traces of hermaphroditism ('pera que com tudo cuidado visem e examinasem se o Reo era mulher ou hermafrodito ou perpetuo ou ad tempos' (in order to investigate and examine whether the accused was a woman or a hermaphrodite, either permanent or temporary)).[127]

This second examination definitively discarded the possibility of Furtado being either a woman or a hermaphrodite and placed the authority of the medical doctors, now restored, over and above that of the witnesses' statements. The record of the trial states the following: the physical examinations 'fazerem mais fee porque nelhas não pode aver falsidade nem emgano o que nas testemunhas pode aver e asim ficãm dibilitadas seus ditos por ser a proua das vestorias' (were more trustworthy because they cannot contain falsehood or error, unlike the testimonies of the witnesses whose claims can be invalidated by the proof supplied by the examinations).[128] As a result of this conclusion, it was deemed that Furtado was male, had not committed sodomy, but had committed the vice of 'luxuria', same-sex masturbation or intercrural sex, and was not any form of hermaphrodite. Finally, it was stated that 'o Diabo o não podia transformar de homem em molher' (the Devil was not capable of transforming him as a man into a woman).[129] This last decision was taken in the light of certain evidence to the contrary; however, the majority of authorities on the subject did not believe such a transformation was possible.

Father Furtado was condemned for deceiving his flock and for possible heretical error, a relatively light sentence given the implications of hermaphroditism, a pact with the devil and sodomy. What the case shows, nevertheless, is that despite the fact that the authorities that entertained the possible existence of hermaphroditism and sex change (by means of diabolical powers) were unfortunately not recorded in the trial papers, inquisitors were prepared at least to entertain the possibility of such phenomena in the case of Furtado.[130] Although the powers of testimonies and expert physical examinations were placed in contest, with the eventual triumph of the latter, Furtado's case shows how competing knowledges were in operation at a time of growing secularization in the scientific sphere, whereby most of those Portuguese observers of hermaphroditism, despite believing in its existence, explained it from a physical bodily perspective. In essence, in many ways, the trial of Furtado was a modern story concerned more with the appropriate behaviour of an ordained priest and his role in guiding his parishioners than with identifying hermaphroditism or the supernatural causes that allowed a man to change into a woman.

A set of religious concerns also played a large part in the final case considered in this section of the chapter. A quarter of a century after the trial of Father Furtado, accusations of sodomy, hermaphroditism and diabolical sex change all circled around José Martins, a shepherd from Ervedal in the Alentejo who was known in the area as a 'macho femea' (she-man).[131] The case came to the attention of the Inquisition in Évora as a result of the concerns raised by the village priest, Father Pedro de São Boavista, in respect of the behaviour and possible sex of one of his parishioners, José Martins. In addition to being remiss in attending mass, Martins had not confessed in Ervedal but in the neighbouring village of Figueira. Furthermore, São Boavista declared that Martins was a 'she-man'. Right from the start, the lack of religious observance was intricately intertwined with the accusation of Martins having had sexual relations with men from whom s/he had borne children.[132] Numerous witnesses declared that Martins had woven cloth on Sundays and festival days, had eaten meat on prohibited days, and had lain with numerous men. The question of sodomy was thus posed, not least because Martins dressed as a man ('vive em trago de homem') and because he 'tem actos carnais com homens fazendo oficio de molher' (performed carnal acts with men, acting as the woman).[133] Despite dressing as a man and displaying a beard, the Inquisition ordered a physical examination of Martins to determine his state. Despite the witness accounts that he was a 'macho-femia', the three medical experts who examined Martins declared that 'lhe virão todas as demostrações de varão' (saw in him all the physical attributes of a man).[134] There was no evidence of a 'vazo femenino' (female vessel), he had plenty of hair on this chest, and his chest was flat 'como de homem' (just like a man).[135] José Martins was therefore declared to be a man. This led to the suspicion of possible sodomy of the passive variety.

Before arriving at this conclusion, however, just as in the case of Father Furtado, the power of the witness statements once again placed the authority of the medical examination in jeopardy. The inquisitors sought clarification from their superiors in Lisbon in light of the continued assertions by witnesses that they were in fact dealing with a hermaphrodite, for Martins had stated that he had had sexual liaisons by means of his 'vazo natural' (natural vessel). As Soyer points out, this expression caused much confusion, but it seemed to undermine any allegation that he had performed sodomy as a man with other men (whereby he would probably have spoken of his 'vazo trazeiro' (rear vessel)). Was Martins thus saying that he was, in fact, a woman, or was he trying precisely to hide any sodomitical tendencies? It is worth quoting fully the meditations on this possibility:

> E como tam era factível que as testemunhas naturalmente pudessem ter tido os dittos actos de copula carnal como com mulher, pello vazo natural; porque nos Ermaphroditos, como dizen os A.A. custumão haver ambos os sexos, assim masculino, com femenino, ou podem tambem haver soo o sexo femenino, e o masculino occulto, mas com aptidão para este; e do mesmo modo que nas criaturas podem haver dous membros simultaneamente como duas Cabeças, e duas maos, assim tambem podem verificarse ambos os sexos, de qualquer dos dous modos.

(As such it was possible that the witnesses could naturally have undertaken these acts of carnal copulation as if with a woman, in the natural vessel, because hermaphrodites, as the learned authors state, can possess both sexes, one masculine and one feminine, or they can have the feminine alone with the masculine hidden but with this aptitude. In the same way as children can possess two members such as two heads simultaneously and two hands, both sexes can also come to pass in either of the aforementioned manners.)[136]

The public reputation of Martins as a hermaphrodite may have indeed explained the statements that s/he was able to 'uzar do sexo feminino' (use the female sex) and that afterwards he was 'restored' to maleness. This subsequent restoration could have been explained in two ways, according to the inquisitors' discussion: firstly, on the basis of Martin's 'real', that is, anatomical hermaphroditism ('o que custuma succeder naturalmente' (that which tends to happen naturally)). Secondly, his or her hermaphroditism could have arisen as a result of diabolical intervention, as in the case of Father Furtado: 'tambem podia ser por arte e illusão do demonio, occultandolhe o seu membro viril, e que sendo homem parecesse mulher' (it could also be as a result of the art and illusion of the devil by hiding his virile member and making a man appear to be a woman).[137]

In order to clear up Martins's possible hermaphroditism, a second physical examination was ordered. Despite the fact that, for the inquisitors at least, Martins's own self-perception was that he was a man (he asserted that 'nos trages e em tudo o mais mostrava ser homem' (in dress and in everything else it was clear he was a man)), his reputation as a hermaphrodite preceded him, according to several declarations in Ervedal. Martins also displayed knowledge of the existence of creatures with two sexes as spoken about by others; but he was unaware, he told his interrogators, that the devil could intervene in order to change the sex of an individual: it was not possible for 'huma mesma creatura sendo do sexo mascolino, intervendo maleficio se possa ocultar este, fingindosse do sexo femenino' (a creature that is of the male sex by means of a spell to hide this sex and pretend to be female).[138]

As in many such cases, the arguments became somewhat circular, but the eventual decision of the tribunal was that there had been no anal penetration and therefore no sodomy, that there was no diabolical pact transforming his sex, and that, as a result of the two physical examinations, there was no evidence of hermaphroditism. The inquisitors were unimpressed by Martins's lack of church-going, and he was admonished to observe more fully his Catholic duties. Here ended the case.

Conclusion

The cases of Father Furtado, Estêvão Luís and José Martins show how important popular culture was in propelling the Inquisition to act, and also how important the weight of individual testimony was in respect of possible sexual transgression and hermaphroditism, to the extent of overruling, temporarily at least, the medical authority of inquisitorial doctors and surgeons. What was perhaps more important in the two cases, nevertheless, was the question of religious observation. In this respect, both Furtado and Martins had erred: the former in regards to his duty as a man of the cloth and whether he could actually provide the Catholic sacrament as a man, and the latter in respect of his duties in observing religious practice and commitment. It is significant in this sense that Martins was denounced by a New Christian, António da Rosa, as having given birth to a child, and this formed part of the rumour that José was indeed a woman.[139] Father Furtado was responsible for the 'religious indoctrination of converts to Christianity from the Maghreb and Africa', making his role as a priest particularly valuable and one which required an unblemished and incontrovertible history.[140] Some of the allegations against both of these individuals emerged out of community rivalries, a factor that led the inquisitors to doubt them; but on occasion some of these resentments went beyond personal ones to take on a religious significance.[141] In the case of Furtado, the priest had refused a certificate to one family's daughter as an Old Christian, alleging that she was in fact descended from 'notorious' Jews. The resultant humiliation meant that the family viewed Furtado with 'great hatred', hence the accusations against his person.[142] In the case of Estêvão Luís, concerns over the existence of African religions and 'witchcraft' in Portugal provided a motivation for the trial of this seventeenth-century 'sodomite'.

In addition to hermaphroditism (and the accusation of sodomy) acting as a way of placing one's religious and sexual integrity under question, the cases outlined in this chapter show the many understandings that were contained by 'hermaphroditism'. Sex change and dual sex were considered possible within a one-sex model either from a natural perspective, as in the work of philosophers such as Zacuto Lusitano and Rodrigo de Castro, or as a result of possible diabolical intervention, as in the thought of inquisitors. Both systems of thought, however, were under constant strain and were threatened with dissolution. Sex change from man to woman was generally rejected as a possibility at the same time as numerous differences in the bodily make-up of men and women were recorded in medical texts written by Portuguese *conversos* or 'nationals'. Although the Inquisition accepted the existence of hermaphrodites and the possibility of diabolical spells, the cases discussed above from the eighteenth century had begun to reject both as explanations for either sexual or religious transgression. In this sense, the Portuguese medical and theological authorities reflected their Spanish counterparts and opened the way towards the secularizing and biological explanations of hermaphroditism of the nineteenth century.

CONCLUSION

This book has attempted two principal undertakings. Firstly, it has tried to place scientific, theological and, to a lesser degree, although the three cannot be separated, social and cultural discourse on hermaphroditism in Iberia within its own context and frame of reference. By this we mean that the framework of intelligibility for the textual treatment of hermaphroditism in Iberia has been progressively analysed on the basis of the scientific and religious discourses of the time and, especially, within the cultural constraints governing ideas and practices on sex and 'gender'. Throughout, we have sought to avoid any hint of essentialism in our analysis whereby acts, behaviours and discourses are considered in light of current considerations on phenomena such as 'transsexualism', 'homosexuality' and 'gender crossing'. As Foucault has argued, discourse is not complicit with our current knowledge; there is no pre-discursive providence that disposes us towards its 'truths'.[1] Instead, we have chosen to employ and to reveal the meanings of terminology that reflect the concepts and discourses utilized during the three hundred or so years considered here, 1500–1800, which, although they altered perceptibly, cannot be seen as simple 'precursors' to later descriptors or frameworks of understanding. Hence, we have employed and interrogated the meanings of 'hermaphroditism' and 'sex change', the significances that surrounded what it meant to be a man or a woman, and the outward and inward signs that appeared to prove this status to doctors, inquisitors and theologians who examined 'ambiguous' or doubtful cases of sexual identity. Accompanying this analysis of the official and scientific discourse on hermaphroditism was a broad range of cultural and social meanings that were encapsulated by hermaphroditism. The influences of these broader cultural concerns – such as the question of the state organization of the population and natural resources, matters relating to potency and reproduction, and notions of national decline – have been incorporated not merely as adjunct categories but, at many points in our analysis, as some of the principal drivers behind concerns about hermaphrodites and the social, medical and cultural risks that they were seen to embody.

Secondly, this book has emphasized a comparative account that, we hope, goes beyond the more standard coverage of the topic whereby Europe is often constrained by a focus on Britain, France and Germany, with some references to ancient Greece. At its heart, of course, is an analysis of Portugal and Spain, countries, especially the former, left off the European map or generally ignored for any significance they may have for a broader understanding of European culture or, more specifically, the history of sexuality. What, if anything, was notable about the two Iberian countries in respect of the questions posed in this book, and what does the study of discourse on 'hermaphroditism' in Iberia bring to a wider comprehension of the fact of 'ambiguous sex'?

The interest in hermaphroditism in Iberia was certainly not exceptional for the period studied. The number of cases uncovered is comparable to other European countries at the time. What is exceptional, and somewhat paradoxical, is that, in comparison to other countries' production of knowledge on hermaphroditism such as France and Germany, there was in general a lack of transcendence of both the cases studied and the analysis provided beyond the borders of Spain and Portugal. While the exceptional ideas of Matheu i Sanz and Tomás Sánchez – with respect to the possibility of hermaphrodites of possessing and using both sexes at once without sanction, and to take religious vows – did find a reception abroad, for example in Italy,[2] most Spanish authors, despite importing knowledge on the subject into Spain, were not successful in having their thought taken up in the other direction. The paradox, if it really is one, is pertinent, especially to the case of Portuguese writers on hermaphroditism. Here, a different set of dynamics was in operation. In spite of the period of forced unification of the two countries that lasted sixty years (1580–1640), Portuguese authors looked, as their Spanish counterparts did, outside of the peninsula for inspiration on the question of hermaphroditism and, in contrast to most of their Spanish colleagues, perceived a reception and an externalization that far exceeded that of the latter. Portuguese authors provided multiple cross-referencing among themselves to the works of Zacuto, the Castros and Reis Franco, for example, and authors such as Amatus Lusitanus transcended national borders to form a reference point on matters regarding sex change in Europe – the story of Maria Pacheca from Esgueira being a case in point. The transnationalism of this thought on medicine and philosophy generally and hermaphroditism and sex change specifically was in part due to the set of factors identified by Sousa Santos with respect to what he has termed the semi-peripheral nature of Portugal *vis-à-vis* Europe.[3] Portugal's geographical location marginalized the country from European trends, but this was compensated by the longevity of the empire, a factor, as Teixeira has argued, that influenced thought on medicine in Portugal in a constant process of innovation.[4] If we add to this the mobility of the Jewish and *converso* intellectual and cultural currents in the sixteenth and seventeenth centuries, we have a powerful set of explanations that reveal the nature of the 'paradox' we have identified.

In both countries, perhaps especially in Spain, moreover, the influence of the work of the Arabic scholars – the means by which Greek philosophical thought was 'rediscovered' in Europe as it was translated from Arabic to Latin and into the vernacular[5] – meant that understandings of hermaphroditism in Iberia were permeated by a highly pluralistic set of knowledges, from Aristotelian, Galenic and Hippocratic notions through to Roman and Jewish law. The work of Gaspar dos Reis Franco is indicative of the multifaceted debates that ensued between the various different models with respect to the possibility or otherwise of hermaphroditism and change from one sex to the other. In Reis Franco's work, although Aristotle's refusal to admit the existence of hermaphrodites is noted, his thought on sex differences is viewed in a particular way whereby it became compatible with Hippocratic and other notions current at the time. As Joan Cadden and others have argued,[6] the dominance of the Hippocratic model – whereby humans were not divided into two sexes, male and female, but one sex, the male sex, with derivations of varying perfection from this one sex – supposedly in force until the eighteenth century, is not so clear-cut as Laqueur has suggested. Numerous differences between the sexes were observed in seventeenth-century works in Iberia, and although there prevailed the idea that Nature moved from the least perfect towards the greatest perfection (hence sex changes from females to males were accepted as realities), the tradition of theories of generation was 'internally diverse'.[7] On the other hand, as the case of Fernanda Fernández shows, at the end of the eighteenth century, the idea that one could change sex – rejected by Aristotle – was still countenanced in Spain, reflecting the currency of the one-sex model informed by notions of perfection.

This internally diverse scenario, to use the words of Lorraine Daston and Katharine Park once more, is evidenced by the discussions held among the inquisitors of the Holy Office in Spain and Portugal (and further afield, in Brazil), as François Soyer has shown extensively.[8] Although the officers of the Inquisition did not name those whose theories they used to inform their judgements, there was a clear flow of ideas from those who analysed hermaphroditism and, consequently, the religious orders and institutions. The idea that hermaphroditism and sex change were possible was given traction by the Inquisition, although the dimension of demonic intervention was often added into discussions of specific cases and charges.

Discourse on hermaphroditism in Iberia as traced in this book responded to the existence – real or otherwise – of an identifiable set of bodily traits and cultural expectations of sex, as expressed through rank, that were seen to upset the natural order of things. The significance of what was understood by 'hermaphroditism' shifted over time in accordance with changing scientific, theological and cultural diagnoses and served to forge connections with broader threats to bodily and national integrity, perceived sinful activity such as sodomy, and concerns over effeminacy and decline. The moving to the margins of certain elements in these relations, when viewed from the perspective of their own genealogy, displays what Foucault has termed the workings of the 'teratology of knowledge';[9] any discourse is always hybrid and is always being reconfigured. The monsters themselves that wander through the history of this knowledge, hermaphroditic in their make-up, return to scare us as times move on. The Portuguese medico-legal expert A. J. Pires de Lima's own investigations in the 1930s, for example,[10] show the shedding of the most monstrous conceptualizations of hermaphroditism – by drawing on the self-same sources that produced the hermaphrodite in the sixteenth century – in order to present new stories about the teratology of the present. But is the object of our knowledge the same thing across these centuries? Pires de Lima clearly believed that his work followed on from the early modern trajectory of the authors he cited. But the 'order of the discourse' was another, to the extent that although certain proximities can be found between the hermaphrodite of the 1600s and that of the early 1800s or early 1900s,[11] we cannot talk of the 'fact' of their sameness but rather of the production of knowledge out of what were, in Ortega's words,[12] radically different 'situations' across the Iberian nations.

NOTES

Introduction: Sex, Gender and Historicity

1. A. A. de Aguiar, 'Pseudo-hermafroditismo feminino (caso português do século XVII)', *Archivo de Medicina Legal*, 2:4 supplement, 1923–5 (1928), pp. 432–6. In Chapter 4, this case and other work by de Aguiar is discussed more fully. A neglected figure in Portuguese medical history, de Aguiar has been the subject of a short account by F. Molina Artaloytia, 'Estigma e interacción: un análisis filosófico del discurso del Dr. Asdrúbal d' Aguiar sobre el homoerotismo', in A. L. Pereira and J. R. Pita (eds), *III Jornadas de História da Psiquiatria e Saúde Mental. Reunião Internacional* (Coimbra: Universidade de Coimbra-CEIS 20, 2012), pp. 7–12.

2. De Aguiar, 'Pseudo-hermafroditismo feminino', p. 432.

3. See, for example, J. A. Pires de Lima, 'Hermafroditismo e Inter-Sexualidade', *A Medicina Contemporânea*, 44 (1939), pp. 473–8.

4. The important study by F. Soyer, *Ambiguous Gender in Early Modern Spain and Portugal: Inquisitors, Doctors and the Transgression of Gender Norms* (Leiden and Boston, MA: Brill, 2012), was published as we were writing this book. The emphasis of this author is placed on the analysis of the thought and practice of the Inquisition in Spain and Portugal and less on the cultural, medical and theological paradigms of the period. In addition, Soyer's book discusses a broader range of 'ambiguous gender' patterns than our prime focus on hermaphroditism.

5. See, for example, C. Bravo-Villasante, *La Mujer Vestida de Hombre en el Teatro Español (siglos XVI–XVII)* (1955; Madrid: Mayo de Oro, 1988); S. Velasco, *The Lieutenant Nun: Transgenderism, Lesbian Desire and Catalina de Erauso* (Austin, TX: University of Texas Press, 2000). Cf. n. 16 for a fuller set of references.

6. See the otherwise excellent K. P. Long, *Hermaphrodites in Renaissance Europe* (Aldershot: Ashgate, 2006) as an example.

7. Portugal was, of course, placed under Spanish rule from 1580, just after the death of the heirless King Sebastian, until the victory of the Portuguese in the war of independence in 1640.

8. M. Foucault, *The History of sexuality, Volume 1: An Introduction* (Harmondsworth: Penguin, 1990), p. 101.

9. See I. Burshatin, 'Written on the Body: Slave or Hermaphrodite in Sixteenth-Century Spain', in J. Blackmore and G. S. Hutcheson (eds), *Queer Iberia: Sexualities, Cultures and Crossings from the Middle Ages to the Renaissance* (Durham, NC: Duke University Press, 1999), pp. 420–56, esp. pp. 423–4. For a later period, but in comparative mode, see F. Molina Artaloytia, 'Los avatares (ibéricos) de la noción de sodomía entre

la Ilustración y el Romanticismo', in F. Durán López (ed.), *Obscenidad, vergüenza, tabú: contornos y retornos de lo reprimido entre los siglos XVIII y XIX* (Cadiz: Universidad de Cádiz, 2012), pp. 101–20.

10. N. J. Efron, 'Nature, Human Nature, and Jewish Nature in Early Modern Europe', *Science in Context*, 15:1 (2002), pp. 29–49, has argued that the marginal or persecuted dimension of Jewish experience contributed to different concepts of nature. On the 'rediscovery' of Greek philosophical thought via the Islamic presence in Italy and Spain in the early new millennium, and the thought of Albucasis on hermaphroditism as presented in his *On Surgery and Instruments*, see R. Cleminson and F. Vázquez García, *Hermaphroditism, Medical Science and Sexual Identity in Spain, 1850–1960* (Cardiff: University of Wales Press, 2009), p. 23 n. 30.

11. H. Kamen, *The Disinherited: The Exiles Who Created Spanish Culture* (London: Penguin, 2008), p. 28.

12. Ibid., p. 30.

13. L. Febvre, 'History and Psychology' (1938), cited in P. Burke, *Varieties of Cultural History* (Ithaca, NY: Cornell University Press, 1997), p. 169. This does not mean that we engage in a 'history of mentality' type study in this book. For the attractiveness but also the weaknesses of this method, see ibid., pp. 162–82.

14. P. Bourdieu, 'Les conditions sociales de la circulation internationale des idées', *Actes de la Recherche en Sciences Sociales*, 145 (2002), pp. 3–8.

15. On exclusion and exteriority, see J. Butler, *Bodies that Matter: On the Discursive Limits of 'Sex'* (New York and London: Routledge, 1993), pp. 11–12.

16. What historians have achieved in this field in the last thirty years can be judged by recalling the words of Jacques Revel in 1982: 'it is not at all surprising if in societies that are libertarian and conflictual at the same time such as our own, the emblematic figure, the symbol of the hermaphrodite, accrues a new significance. The social and cultural function of these social representations of the intermediate is still to be studied, however'; see J. Revel, 'El historiador y los papeles sexuales', in Various Authors, *Familia y Sexualidad en Nueva España* (Mexico: FCE, 1982), pp. 53–4. For *Ancien Régime* Spain, see F. Vázquez García and A. Moreno Mengíbar, 'Un solo sexo. Invención de la monosexualidad y expulsión del hermafroditismo', *Daimón. Revista de Filosofía*, 11 (1995), pp. 95–112; A. Morel d'Arleux, 'Las "Relaciones de Hermafroditas": dos ejemplos diferentes de una misma manipulación ideológica', in M. Cruz García de Enterría, H. Ettinghausen, V. Infantes and A. Redondo (eds), *Las Relaciones de sucesos en España (1500–1750). Actas del Primer Coloquio Internacional (Alcalá de Henares, 8, 9 y 10 de junio de 1995)* (Alcalá de Henares and Paris: Pub. Universidad Alcalá de Henares/Pub. Sorbonne, 1996), pp. 261–71; F. Vázquez García and A. Moreno Mengíbar, *Sexo y Razón. Una genealogía de la moral sexual en España (siglos XVI–XX)* (Madrid: Akal, 1997), pp. 185–204; F. de la Flor, 'La "puella pilosa". Representaciones de la alteridad femenina (de Sánchez Cotán a José de Ribera, pasando por Sebastián de Covarrubias)', in *La Península Metafísica. Arte, literatura y Pensamiento en la España de la Contrarreforma* (Madrid: Biblioteca Nueva, 1999), pp. 267–305; F. Vazquez García and A. Moreno Mengíbar, 'Hermafroditas y cambios de sexo en la España Moderna', in A. Lafuente and J. Moscoso (eds), *Monstruos y Seres Imaginarios en la Biblioteca Nacional* (Madrid: Ministerio de Educación y Cultura, Biblioteca Nacional, 2000), pp. 91–103; M. Cátedra, 'Sobre la ambigüedad: el caso de Paula Barbada', in E. Crespo and C. Soldevilla (eds), *La Constitución Social de la Subjetividad* (Madrid: Los Libros de La Catarata, 2001), pp. 131–44; S. Velasco, 'Marimachos, hombrunas, barbados: The Masculine Woman in Cervantes', *Cervantes: Bulletin of the*

Cervantes Society of America, 20:1 (2001), pp. 69–78; V. Marchetti, *L'invenzione della bisessualità. Discussioni tra teologi, medici e giuristi del XVII secolo sull'ambiguità dei corpi e delle anime* (Milan: Mondadori, 2001), pp. 217–66; M. J. de la Pascua Sánchez, '¿Hombres vueltos del revés? Una historia sobre la construcción de la identidad sexual en el siglo XVIII', in M. J. de la Pascua Sánchez, M. del Rosario García Doncel and G. Espigado (eds), *Mujer y Deseo* (Cadiz: Universidad de Cádiz, 2003), pp. 431–44; E. del Río Parra, *Una era de monstruos. Representaciones de lo deforme en el Siglo de Oro español* (Madrid: Iberoamericana, 2003), pp. 86–100; S. Velasco, *Male Delivery: Reproduction, Effeminacy and Pregnant Men in Early Modern Spain* (Nashville, TN: Vanderbilt University Press, 2006); A. Salamanca Ballesteros, *Monstruos, Ostentos y Hermafroditas* (Granada: Universidad de Granada, 2007); M. J. Zamora Calvo, '*In virum mutata est*. Transexualidad en la Europa de los siglos XVI y XVII', *Bulletin Hispanique*, 110:2 (2008), pp. 431–47; F. Soyer, 'The Inquisition and the "Priestess of Zafra": Hermaphroditism and Gender Transgression in Seventeenth-Century Spain', *Annali della Scuola Normale Superiore di Pisa. Classe di Lettere e Filosofia*, 5:1–2 (2009), pp. 535–62; and M. Alcalá Galán, 'El andrógino de Francisco de Lugo y Ávila: discurso científico y ambigüedad erótica', *e-Humanista*, 15 (2010), pp. 107–35. Marchetti, *L'invenzione della bisessualità*, has explored in minute detail the role of Spanish authorities in these matters and their impact on Europe more broadly.

17. The first publication by Foucault on the question of Herculine Barbin was M. Foucault, *Herculine Barbin dite Alexina B.* (Paris: Gallimard, 1978) on the basis of a reading of a psychiatric report from the mid-nineteenth century. Foucault's idea was to write up a number of cases like Barbin's to include in one of the volumes of his history of sexuality (cf. 'Chronologie', in M. Foucault, *Dits et Écrits 1954–1988*, vol. 1 (Paris: Gallimard, 1994), p. 54). He discusses cases from the sixteenth to the eighteenth century in M. Foucault, *Les Anormaux. Cours au Collège de France 1974–1975* (Paris: Gallimard-Le Seuil, 1999), pp. 51–74. The first edition of the book by Thomas Laqueur is *Making Sex: Body and Gender from the Greeks to Freud* (Cambridge, MA and London: Harvard University Press, 1990).

18. J. Butler, *Gender Trouble: Feminism and the Subversion of Identity* (New York: Routledge, 1990), pp. 119–41.

19. The historical analysis by Foucault has been critiqued by several historians (e.g. Katharine Park, Thomas Laqueur and Ruth Gilbert) and by some philosophers (e.g. Judith Butler) for his tendency to idealize certain historical facts (the freedom of hermaphrodites to adopt a particular sex in societies under the *Ancien Régime*) or certain aspects related to sexual life (the pleasures enjoyed by Herculine Barbin, apparently unconnected to the prevailing constraints of identity). A good summary of the debate in the mid-1990s is C. J. Nederman and J. True, 'The Third Sex: The Idea of the Hermaphrodite in Twelfth-Century Europe', *Journal of the History of Sexuality*, 6:4 (1996), pp. 497–517, on p. 498 n. 5. Among the first critiques of the thesis by Laqueur are Joan Cadden, Sally Shuttleworth, Peter Laipson, Katharine Park, Lorrain Daston, Robert Nye, Patricia Parker and Gianna Pomata. Those who argue in favour of Laqueur's ideas include Ann Rosalind Jones, Peter Greenblatt, Peter Stallybrass and Londa Schiebinger who apply his thesis to their work. A second stage of the debate took off in 2003, as reflected in *Isis*. This was a three-way debate between Laqueur himself, Michael Stolberg and Schiebinger. Joan Cadden intervened once more in 2004. In the first stage of this controversy, Laqueur was accused of proposing a unilateral reading of medical sources in order to highlight the importance of the monist Hippocratic-Galenic model over and above the dualist Aristotelian model. In the second stage, led by Stolberg, the displacement of the monist model

by the dualist model was accepted to have taken place, but only in Renaissance medicine between the sixteenth and seventeenth centuries. Cf. M. Stolberg, 'A Woman Down to her Bones: The Anatomy of Sexual Difference in the Sixteenth and Early Seventeenth Centuries', *Isis*, 94:2 (2003), pp. 274–99; L. Schiebinger, 'Skelettestreit', *Isis*, 94:2 (2003), pp. 307–13; and T. Laqueur, 'Sex in the Flesh', *Isis*, 94:2 (2003), pp. 300–6.

20. We use this concept in the same way as Passeron, who considers history and sociology as epistemologically equivalent disciplines. See J. C. Passeron, *Le Raisonnement Sociologique. Un espace non-poppérien de l'argumentation* (Paris: Albin Michel, 2006), pp. 125–68.

21. Two examples of this kind of error in studies of the 'hermaphrodite' Elena de Céspedes in the sixteenth century would be M. Escamilla, 'A propos d'un dossier inquisitorial des environs de 1590: les étranges amours d'un hermaphrodite', in A. Redondo (ed.), *Amours légitimes, Amours illégitimes en Espagne (XVIe–XVIIe Siècles)* (Paris: Pub. de la Sorbonne, 1985), pp. 167–82, and E. Maganto Pavón, *El proceso inquisitorial contra Elena/o de Céspedes (1587–1588) (Biografía de una cirujana transexual del siglo XVI)* (Madrid: Método Gráfico, 2007).

22. This distinction between discursive levels has been used in F. Vázquez García, 'Del Hermafrodita al Transexual. Elementos para una genealogía del cuerpo sexuado (España siglos XVI–XX)', in N. Corral (ed.), *Prosa Corporal. Variaciones sobre el cuerpo y sus destinos II* (Madrid: Talasa, 2008), pp. 75–97. This perspective concurrs with that elaborated by Butler.

23. See Salamanca Ballesteros, *Monstruos*, pp. 283–312, where the past is interrogated on the basis of the concepts present precisely in their own historical period.

24. M. Foucault, 'Le vrai sexe', in *Dits et Écrits 1954–1988*, vol. 4 (Paris: Gallimard, 1994) (orig. 1980), pp. 115–23, on p. 116.

25. See Laqueur, *Making Sex*, p. 124; K. Park, 'The Rediscovery of the Clitoris: French Medicine and the Tribade, 1570–1620', in D. Hillman and C. Mazzio (eds), *The Body in Parts: Fantasies on Corporeality in Early Modern Europe* (New York and London: Routledge, 1997), pp. 171–93, on p. 174; Nederman and True, 'The Third Sex', p. 516; and R. Gilbert, *Early Modern Hermaphrodites* (Basingstoke: Palgrave, 2002), pp. 2–3. Despite these critiques, some historians still uncritically adopt Foucault's perspective, e.g. B. Enguix, 'Cuerpo y transgresión: de Helena de Céspedes a Lady Gaga', *Revista Latinoamericana de Estudios sobre Cuerpos, Emociones y Sociedad*, 5:3 (2011), pp. 25–38, on p. 30.

26. This representation of the hermaphrodite as a 'third sex' is what Nederman and True consider to be prevalent in the Christian West in the twelfth century (Nederman and True, 'The Third Sex'). The hypothesis of the third sex as a transcultural phenomenon has been defended by some anthropologists such as G. Herdt (ed.), *Third Sex, Third Gender: Beyond Sexual Dimorphism in Culture and History* (New York: Zone Books, 1994). A. Fausto-Sterling, *Sexing the Body. Gender Politics and the Construction of Sexuality* (London: Basic Books, 2000), pp. 78–114, argues for at least five sexes in the human species.

27. See, for example, the ways in which Laqueur minimizes the differences of Aristotelianism with respect to the 'one-sex' model (Laqueur, *Making Sex*, pp. 28–33) and how he tries to present as support for his interpretation the medical belief in 'menstruating men' (Laqueur, 'Sex in the Flesh', p. 305). While Laqueur argues that in the one-sex model the male body is the paradigm from which the female follows, for Gianna Pomata the existence of menstruating males would imply the contrary interpretation. See G. Pomata, 'Uomini Menstruanti. Somiglianza e Diferenza Fra i Sessi in Europa in Etá Moderna', *Quaderni Storici*, 27:1 (1992), pp. 51–103, on p. 56. For his part, Michael Stolberg appears to maintain the thesis that the single-sex model was replaced by the dual-sex

model but argues that this transformation in fact took place later, between the sixteenth and the seventeenth centuries. See Stolberg, 'A Woman Down to her Bones'.

28. J. Cadden, *Meanings of Sex Difference in the Middle Ages: Medicine, Science, and Culture* (Cambridge: Cambridge University Press, 1993), p. 3. A further example of resistance to 'heuristic simplifications' can be found in Park, 'The Rediscovery of the Clitoris', p. 174.

29. Cadden, *Meanings of Sex Difference*.

30. L. Daston and K. Park, 'The Hermaphrodite and the Orders of Nature: Sexual Ambiguity in Early Modern France', *GLQ: A Journal of Gay and Lesbian Studies*, 1:4 (1995), pp. 420–5; L. Daston and K. Park, 'The Hermaphrodite and the Orders of Nature: Sexual Ambiguity in Early Modern France', in L. Fradenburg and C. Freccero (eds), *Premodern Sexualities* (New York and London: Routledge, 1996), pp. 117–36; Park, 'The Rediscovery of the Clitoris', pp. 179–87; L. Daston and K. Park, *Wonders and the Order of Nature, 1150–1750* (New York: Zone Books, 1998), p. 203.

31. Marchetti, *L'invenzione della bisessualità*, p. 316.

32. S. Greenblatt, *Shakespearean Negotiations: The Circulation of Social Energy in Renaissance England* (Oxford: Clarendon Press, 2001), pp. 88–91, points out that in Renaissance culture identity is the result of a process teleologically oriented by the masculine default position.

33. G. Deleuze, *The Fold: Leibniz and the Baroque*, trans. T. Conley (Minneapolis, MN: University of Minnesota Press, 1993), pp. 64–7, identifies individual difference as part of Baroque culture as a 'fold' inserted inside a continuum in society.

34. In a somewhat confusing manner, Elisabeth Perry seems to suggest that at this time biology was considered to constitute a fixed destiny as there was no concept of 'gender': 'most people of this time lacked any concept of a socially constructed gender, and most believed in sex as an essential quality granted at birth, integral to the "natural order", and essential to a sexually dichotomized, hierarchical and patriarchal sociopolitical system'. See M. E. Perry, 'From Convent to Battlefield: Cross-Dressing and Gendering the Self in the New World of Imperial Spain', in J. Blackmore and G. S. Hutcheson (eds), *Queer Iberia: Sexualities, Cultures and Crossings from the Middle Ages to the Renaissance* (Durham, NC: Duke University Press, 1999), pp. 394–419, on p. 411.

35. On Catalina de Erauso, see Perry, 'From Convent to Battlefield'; Velasco, *The Lieutenant Nun*; and N. Aresti Esteban, 'The Gendered Identities of the "Lieutenant Nun": Rethinking the Story of a Female Warrior in Early Modern Spain', *Gender and History*, 19:3 (2007), pp. 401–18. On Elena de Céspedes, see M. C. Barbazza, 'Un caso de Subversión Social: el proceso de Elena de Céspedes (1587–1589)', *Criticón*, 26 (1984), pp. 17–40; Vázquez García and Moreno Mengíbar, 'Un solo sexo', pp. 99–103; Burshatin, 'Written on the Body'; and R. Kagan and A. Dyer, *Inquisitorial Inquiries: Brief lives of Secret Jews and Other Heretics* (Baltimore, MD: Johns Hopkins University Press, 2004), pp. 36–59. On Fernanda Fernández, see de la Pascua Sánchez, '¿Hombres vueltos al revés?'

36. We have termed this complex the 'regime of true rank' in Vázquez García, 'Del Hermafrodita al Transexual'. Although we differ from some aspects of the methodology and theoretical framework utilized by V. L. Bullough and B. Bullough in their *Cross Dressing, Sex, and Gender* (Philadelphia, PA: University of Pennsylvania Press, 1993), pp. 45–7, we coincide with their useful comment on the question of status in the Middle Ages.

37. This arrangement is discussed by Laqueur, *Making Sex*, pp. 135 and 137–8. On clothing as distinctive of rank in this society, see J. Lalinde Abadía, 'La indumentaria como símbolo de discriminación jurídico social', *Anuario de Historia del Derecho Español*, 53 (1986), pp. 583–601.

38. The otherwise excellent piece by de la Pascua Sánchez on the case of María Fernández at the end of the eighteenth century falls into this trap: 'in the story I cover here, the biological elements overlap and desire becomes the protagonist and guide in the construction of a new sexual identity. This fact implies modernity ... because the body appears not as something "given" naturally but as something explored as a result of sexual desire' (de la Pascua Sánchez, '¿Hombres vueltos al revés?', p. 439). Long, *Hermaphrodites in Renaissance Europe*, pp. 1 and 243, falls into this error when she suggests fluidity of sex and gender in the Renaissance by drawing on the work of Dollimore, Haraway and Butler. This 'postmodern' inflection, which has for good reason been designated 'neo-Baroque' in its consideration of sexual metamorphoses such as transvestism and drag, deconstructs a sex/gender division that has no sense in the modern period. On the neo-Baroque condition of questions on 'trans' and the postmodern, see O. Calabrese, 'Neobarroco', in F. Jarauta (ed.), *Otra Mirada sobre la Época* (Murcia: Colegio Oficial de Aparejadores y Arquitectos Técnicos/Cajamurcia, 1994), pp. 261–2.

39. This does not mean that we reject entirely the category 'gender' when analysing transgressions against 'sex as status' in this period. But 'gender', in inverted commas, would have to be understood in the sense employed by Judith Butler, that is, as something that includes 'sex' and 'sexuality' as constituted by *body performances*. 'Gender', then, forms part of regimes of historically changing relations. These regimes would include the performative acts and their transgressions in any specific historical period. On this 'resignification' or subversive reiteration of 'gender', see Butler, *Bodies that Matter*, pp. 121–40.

40. The idea of 'bare life' is taken from G. Agamben, *Homo Sacer: Sovereign Power and Bare Life*, trans. D. Heller-Roazen (Stanford, CA: Stanford University Press, 1998).

41. Greenblatt, *Shakespearean Negotiations*, p. 179.

42. F. Vázquez García, 'La imposible fusión. Claves para una genealogía del cuerpo andrógino', in D. Romero de Solís, J. B. Díaz-Urmeneta Muñoz and J. López-Lloret (eds), *Variaciones sobre el cuerpo* (Seville: Servicio de Publicaciones de la Universidad de Sevilla, 1999), pp. 217–35, on pp. 222–3.

43. Laqueur, *Making Sex*, p. 149.

44. Cf. R. Espósito, *Bíos. Biopolítica y Filosofía* (Buenos Aires: Amorrortu, 2006), pp. 88–9.

45. Foucault, *The History of Sexuality, Volume 1*, p. 121.

46. N. Elias, *The Court Society*, trans. E. Jephcott (Oxford: Blackwell, 1983), p. 255.

47. Ibid., p. 256, and N. Zemon Davis, *Il Ritorno di Martin Guerre. Un caso di doppia identità nella Francia del Cinquecento* (Turin: Einaudi, 1984), p. 58.

48. R. Castel, *Les métamorphoses de la question sociale: une chronique du salariat* (Paris: Fayard, 1995), pp. 40–89.

49. For Spanish examples, see R. Martínez, 'Mari(c)ones, travestis y embrujados. La heterodoxia del varón como recurso cómico en el teatro breve del Barroco', *Anagórisis*, 3 (2011), pp. 9–37, on pp. 14–15.

50. Greenblatt, *Shakespearean Negotiations*, p. 88.

51. Bravo-Villasante, *La Mujer Vestida de Hombre*, pp. 15–33 and 78. See also G. Bradbury, 'Irregular Sexuality in the Spanish "Comedia"', *Modern Language Review*, 76:3 (1981), pp. 566–80, and M. McKendrick, *Woman and Society in the Spanish Drama of the Golden Age: A Study of the 'Mujer Varonil'* (Cambridge: Cambridge University Press, 1974).

52. Greenblatt, *Shakespearean Negotiations*, p. 92.

53. See Laqueur, *Making Sex*, pp. 136–7, on the deliberations of the judges on the case of Marin le Marcis at the beginning of the seventeenth century. The definitive work on juridical controversies in this sense is that of Marchetti, *L'invenzione della bisessualità*.

On the notion of *praevalet*, or predominant sex, see Marchetti, *L'invenzione della bisessualità*, pp. 67–9. J. Rappaport, 'Mischievous Lovers, Hidden Moors and Cross-Dressers: Passing in Colonial Bogotá', *Journal of Spanish Cultural Studies*, 10:1 (2009), pp. 7–25, has asserted that 'race' was likewise not a recognizable and purely biological category because it formed part of what was understood to be 'quality', that is, part of the regime of rank.

54. Vázquez García, 'Del Hermafrodita al Transexual'.
55. M. Bakhtin, *La Cultura Popular en la Edad Media y en el Renacimiento* (Madrid: Alianza Universidad, 1987), pp. 273–331.
56. Gilbert, *Early Modern Hermaphrodites*, pp. 2, 5 and 9.
57. K. Park, 'Una historia de la admiración y del prodigio', in A. Lafuente and J. Moscoso (eds), *Monstruos y Seres Imaginarios en la Bilbioteca Nacional* (Madrid: Ministerio de Educación y Cultura, Biblioteca Nacional, 2000), pp. 77–90.
58. Laqueur correctly points out that in the absence of any system that fixed the sexes on a biological basis, institutions attempted to consolidate difference on the basis of severe punishment of any transgression (Laqueur, *Making Sex*, p. 125). However, his work centres on medical accounts, travel literature and stories of marvels, and hardly touches upon sources of sexual ambiguity as a sign of sin or negative portent. On transvestism as a challenge to heteronormativity, see Butler, *Bodies that Matter*, pp. 121–37. On the question of sodomy, mollities and the Catholic stance, see P. Hurteau, 'Catholic Moral Discourse on Male Sodomy and Masturbation in the Seventeenth and Eighteenth Century', *Journal of the History of Sexuality*, 4:1 (1993), pp. 1–26. Specifically on the sodomy–heresy–Inquisition relation, see the recent F. Molina, 'La *herejización* de la sodomía en la sociedad moderna. Consideraciones teológicas y *praxis* inquisitorial', *Hispania Sacra*, 62:126 (2010), pp. 539–62.
59. We take this three-dimensional structure (*mirabilis, magicus, miraculus*) from the work of J. Le Goff, *The Medieval Imagination*, trans. A. Goldhammer (Chicago, IL and London: University of Chicago Press, 1992), pp. 27–44. On the concept of *mirabilis*, cf. the excellent piece by Park, 'Una historia de la admiración y del prodigio'.

1 Marvels, Monsters and Prodigies: Hermaphrodites as Natural Phenomena in Spain, 1500–1700

1. Nederman and True, 'The Third Sex', p. 501; Daston and Park, *Wonders and the Order of Nature*, p. 49; Vázquez García and Moreno Mengíbar, *Sexo y Razón*, pp. 187–8.
2. C. Kappler, *Monstruos, Demonios y Maravillas a fines de la Edad Media* (Madrid: Akal, 1986), pp. 334–5 (French original *Monstres, démons et merveilles à la fin du Moyen Âge* (Paris: Payot, 1980)).
3. De Granada writes of the 'hermosura de las cosas que por la divina Providencia ... fabricadas' (beauty of the things that ... have been fabricated by divine Providence), including the roundness of the earth, the flowers and the trees, which in 'su grande variedad nos son causa de un insaciable gusto y deleite' (their great variety is cause in us of insatiable pleasure and delight); L. de Granada, *Introducción al Símbolo de la Fe*, in *Obras*, vol. 1 (1583; Madrid: BAE, 1944), pp. 81–633, on p. 100.
4. G. Canguilhem, 'La Monstruosité et le Monstrueux', in *La Connaissance de la Vie* (Paris: Vrin, 1980), pp. 178–9, locates the beginning of this process in 'mechanistic physics and philosophy', as a contrast to the bestiaries of the Renaissance. K. Park and L. Daston,

'Unnatural Conceptions: The Study of Monsters in Sixteenth and Seventeenth Century France and England', *Past and Present*, 92 (1981), pp. 20–54, on pp. 36–7, also follow this teleological argument. Later on, in their *Wonders and the Order of Nature*, p. 176, they rectify this error. Gilbert, *Early Modern Hermaphrodites*, pp. 13, 24 and 81, oscillates between a synchronic model and a teleological schema which passes from the Renaissance ideal of the androgyne as a perfect form (in alchemy and neo-Platonism), to the naturalization of the hermaphrodite during the Enlightenment and through to the stigmatization of the figure in the following period.

5.   Daston and Park, *Wonders and the Order of Nature*, p. 175.

6.   The text by A. Gutiérrez de Torres, *El Sumario de las Maravillosas y Espantables Cosas que en el Mundo han acontecido* (1524; Madrid: Real Academia de la Lengua Española, 1952) falls into this category. Despite what the first part of its title might imply, this book refers to ominous prodigies as divine signs of calamities to come. The first part of the book interprets numerous historical events, including some in the history of Spain, in accordance with this perspective. The second part is a kind of treatise on astrology. For this reason, we include it in the field of *magicus* rather than that of *mirabilis*.

7.   Marchetti, *L'invenzione della bisessualità*, p. 37, has referred to the impossibility of establishing clear borders between the different disciplinary genres (canon law, penal law, civil law, medicine, travel books, natural history, demonology and theology) where the problems raised by hermaphrodites and sex change were discussed.

8.   The literature on marvels (*maravillas*) does not exhaust the field of the *mirabilis*. Belonging to this set are also the 'cámaras de maravillas' (from the German *Wunderkammer*). These exhibited, among other things, monsters and portents. In Spain they first found collections during the reign of Felipe II and included, in addition to arms and tapestries, paintings representing these beings, among these the potrait of Brígida de Peñaranda, the famous bearded lady, painted by Sánchez Cotán in 1590. Brígida would exhibit herself in exchange for money. See Salamanca Ballesteros, *Monstruos*, pp. 253–79. On the interest in monstrosity in the court of Spain and that of the Austrias, see F. Bouza, *Locos, enanos y hombres de placer en la Corte de los Austrias* (Madrid: Temas de Hoy, 1991), and A. E. Pérez Sánchez, J. Gallego and M. Mena, *Monstruos, enanos y bufones en la Corte de los Austrias* (Madrid: Fundación Amigos del Museo de Prado, 1986). On the 'curiosities of Nature' included in the 'cámara' of Juan de Lastanosa (1607–84), the *hidalgo* friend of Baltasar Gracián, see E. Correa Calderón, *Baltasar Gracián. Si vida y su obra* (Madrid: Gredos, 1961), pp. 26–32.

9.   The eighteenth-century J. de Arriaga, *Piscator Murciano. Con un agregado de prodigios, cosas no comunes y fuera de el estado natural, que han sucedido, y dignas de que se sepan, como haberse buelto muchas mujeres hombres* (Madrid, 1746), belongs to this genre.

10.   'Todas son cosas que nos admiran, porque no sabemos la orden y causa que llevan; pero ello su razón y sucesso tiene, que Dios lo sabe y ordena' (All these are things that cause admiration in us because we do not know the order and cause that they contain, but reason and existence they have as known and ordered by God); P. Mexía, *Silva de Varia Lección* (1540; Madrid: Cátedra, 1989), p. 502. A. de Torquemada, *Jardín de Flores Curiosas* (1570; Madrid: Castalia, 1982), pp. 105–6, argues the same for the 'máquina y composición del mundo' (machine and composition of the world), the existence of the planets and the sun, and the different types of trees and fruits. For this reason, he says, 'tampoco nos deben de dar causa de maravillarnos, cuando viéremos otras cosas que salgan algún tanto de esta orden tan concertada de la naturaleza' (when we see other things that are somewhat out of the concerted order of nature, neither should we be marvelled by them). This was, in fact, 'Porque ellas no salen ni exceden de naturaleza, que la falta

está en nosotros y en nuestro entendimiento y juicio, que con su torpeza no lo alcanza' (Because they are not outside of and do not exceed nature; the error is in ourselves and in our understanding and judgement, which being limited do not grasp [these phenomena]). A. de Fuentelapeña, *El Ente Dilucidado. Tratado de Monstruos y Fantasmas* (1676; Madrid: Editora Nacional, 1978), p. 80, presented a similar interpretation.

11. Torquemada, *Jardín de Flores Curiosas*, pp. 116–17; on p. 117 Torquemada mentions the authority of Aristotle in order to verify the find. The same reference to the testimony of Calliphanes on the Nasamones appears in S. Covarrubias, 'Ermaphrodito', in *Tesoro de la Lengua Castellana o Española* (1611; Madrid: Turner, 1979), p. 531. The people referred to lived near the Nasamones and were termed the Machlyes. Prodigies and portents described in the *Silva de Varia Lección* (1540) by Pedro de Mexía do not include episodes of sex change. However, two events portraying women who change dress and take on the male roles of pope (the famous Pope Joan) and emperor are mentioned: 'tomando ábitos de hombre, llamándose Juan ... venida desde algunos años en la ciudad de Roma y todavía en hábitos de hombre, tuvo cátedra y enseñó públicamente' (taking the habit of a man, he was called Juan ... came some years ago from the city of Rome and, still dressed as a man, took a chair and taught publicly) (Mexía, *Silva de Varia Lección*, p. 238). 'Amazon' women are also discussed (ibid., pp. 244–61). Finally, the case of Heliogabalus, who tried to operate on himself to become a woman, was also mentioned (ibid., p. 713). The cases of Heliogabalus and Nero as voluntary attempts by men to become women are discussed in a French account of marvels translated in 1603: P. Bovistuau, C. Tesserant and F. Belleforest, *Historias prodigiosas y maravillosas* (Madrid: Imp. Luis Sánchez, 1603).

12. A. de Fuentes, *Summa de Philosophía Natural* (Seville: J. León, 1547), cited in J. Sanz Hermida, 'La literatura de problemas en España (siglos XVI y XVII)' (unpublished doctoral thesis, Universidad de Salamanca, 1997), pp. 1005–6: 'unas mugeres se forman un poco hazia la parte derecha de la madre, que llamamos varoniles; y aquestas son más calientes que no las otras mugeres, pero son menos calientes que los hombres y por esto crían barva. Pero tienen menos barva que los hombres' (some women are formed a little towards the right side of the mother; we call these mannish. These are warmer than other women but not as warm as men and this is why they grow beards. But they have less beard than men).

13. On the 'natural' condition of hermaphrodites and sex changes, see Torquemada, *Jardín de Flores Curiosas*, pp. 116 and 187–90, and Fuentelapeña, *El Ente Dilucidado*, pp. 181 and 242. On changes in sex as natural events, see J. Pérez de Moya, *Philosophía Secreta* (1585; Madrid: Francisco Sánchez, 1628), fol. 263; J. de Pineda, *Treinta y Cinco Diálogos Familiares de Agricultura Cristiana* (Salamanca: Pedro de Adurça y Diego López, 1589), fol. 109r; J. de la Cerda, *Libro intitulado vida política de todos los estados de mujeres* (Alcalá de Henares: Juan Gracián, 1599), p. 518; Martín del Río, *La Magia Demoníaca. Libro II de las Disquisiciones Mágicas* (1599–1600; Madrid: Hiperión, 1991), p. 395; J. E. Nieremberg, *Curiosa y Oculta Filosofía* (1638; Madrid: Imprenta Real, 1643), pp. 54–5; and F. de Torreblanca y Villalpando, *Epithomes delictorum sive de Magia* (Lyon: Juan Antonio Huguet, 1678), book 2, ch. 17. Fuentelapeña also admits the existence of menstruating men, although he explains them by stating that they are in reality 'hidden hermaphrodites' who have both natures, one internal and the other external (Fuentelapeña, *El Ente Dilucidado*, p. 230). The texts by Martín del Río and de Torreblanca y Villalpando can be considered treatises on demonology. However, their distinctiveness from the literature of marvels is unclear in many respects. The treatises by Torquemada and Fuentelapeña also include chapters on witchcraft, sorcery and diabolical intervention.

14. Martín del Río, *La Magia Demoníaca*, p. 393, and de la Cerda, *Libro intitulado vida política de todos los estados de mujeres*, p. 519.
15. Torquemada, *Jardín de Flores Curiosas*, p. 190.
16. Padre Nieremberg also refers to the action of imagination for the change of sex in Nieremberg, *Curiosa y Oculta Filosofía*, pp. 54–5. The imagination of parents was often invoked as a cause of monstrous births, as Fuentelapeña, *El Ente Dilucidado*, pp. 170–1, pointed out. On this topic and the influence of the imagination on births, a Galenic idea principally disseminated by Avicenna to Western medicine, see J. González Rovira, 'Imaginativa y nacimientos prodigiosos en algunos textos del Barroco', *Criticón*, 69 (1997), pp. 21–31.
17. Alcalá Galán, 'El andrógino', p. 111.
18. The error that Zamora Calvo, '*In virum mutata est*', p. 433, in our view, makes.
19. 'Otra cosa es dignísima de ser notada, que ningún miembro tiene el varón que no le tenga la hembra, sino que en el varón están algunos descubiertos, que en la hembra están ocultos, por ser de más frío temperamento, por lo cual parece haber dicho Aristóteles que es macho ocasionado o menguado que es decir que por pequeña ocasión dejó de ser macho' (Another thing is worthy of note: that the male has no organ that the female does not have; in the male some of these are uncovered and in the female hidden as a result of her temperament. For this reason, Aristotle has observed that the female is a failed or reduced man, which means that as a result of some small event she failed to become a man) (Pineda, *Treinta y Cinco Diálogos*, fol. 163v).
20. Martín del Río wrote that Galen and Avicenna and others believed that 'la mujer es una especie de varón monstruoso e imperfecto (woman is a kind of imperfect and monstrous male) (Martín del Río, *La Magia Demoníaca*, p. 395); Fuentelapeña, *El Ente Dilucidado*, p. 244, notes that 'es más frecuente la mutación de muger en hombre que la de hombre en muger … porque la naturaleza siempre aspira a lo más pefecto' (the mutation of woman towards man is more common than man towards woman … because nature always aspires towards the most perfect). Pineda, *Treinta y Cinco Diálogos*, fol. 109r, discounted stories of men becoming women.
21. Stolberg, 'A Woman Down to her Bones'. Laqueur, 'Sex in the Flesh', pp. 301–2, has qualified Stolberg's position as belonging to the dualist understanding of sex.
22. Fuentelapeña agrees with Aristotle that the hermaphrodite is a monster, but not in the same way. Instead of referring to a deviation or sin of Nature, Fuentelapeña prefers to refer to the hermaphrodite as 'extraviarse de lo ordinario' (moving away from the ordinary) (Fuentelapeña, *El Ente Dilucidado*, p. 181). For this reason, the hermaphrodite ('sexo hermafrodítico') with two sexes is not as monstrous 'como el sexo que llaman neutro', i.e. the undefined sex.
23. Fuentelapeña describes generation as a battle between male and female semen; the hermaphrodite results from an inconclusive battle in which neither 'pueda vencer y consumir a la otra' (is able to defeat and consume the other) (ibid., p. 181). He even believes that the hermaphrodite can fertilize itself (ibid., pp. 229–41).
24. Ibid., pp. 247–8.
25. Ibid., p. 248.
26. Del Río, *La Magia Demoníaca*, p. 395.
27. Ibid.
28. Ibid.
29. Ibid., p. 196.
30. Ibid.

31. The *Relación* dates from 1617, although the event must have taken place before 1613 (cf. Morel d'Arleux, 'Las "Relaciones de Hermafroditas"', p. 268).

32. Cf. H. Ettinghausen (ed.), *Noticias del Siglo XVII: Relaciones Españolas de Sucesos Naturales y Sobrenaturales* (Barcelona: Puvill Libros, 1995), p. 12. On this genre of text, see M. Cruz García de Enterría, H. Ettinghausen, V. Infantes and A. Redondo (eds), *Las Relaciones de sucesos en España (1500–1750). Actas del Primer Coloquio Internacional (Alcalá de Henares, 8, 9 y 10 de junio de 1995)* (Alcalá de Henares and Paris: Pub. Universidad Alcalá de Henares/Pub. Sorbonne, 1996).

33. On the *relaciones* of prodigies as an instrument of social control wielded by the Church and authorities, see Ettinghausen, *Noticias del Siglo XVII*, p. 14, and H. Ettinghausen, 'Sexo y Violencia: noticias sensacionalistas en la prensa española del siglo XVII', *Edad de Oro*, 12 (1993), pp. 95–107.

34. Facsimile edition of the *relación* in Ettinghausen, *Noticias del Siglo XVII*, n.p., with a resume on pp. 36–7. The case is commented upon extensively by Morel d'Arleux, 'Las "Relaciones de Hermafroditas"', pp. 263–8.

35. The accompanying symbolism is examined minutely by Morel d'Arleux, 'Las "Relaciones de Hermafroditas"', pp. 265–6. Ettinghausen also suggests the allusion to hermaphroditism but interprets the case in accordance with modern scientific language. He writes: 'it would seem to be a case of hermaphroditism or rather of mixed gonadal disgenesis' (Ettinghausen, *Noticias del Siglo XVII*, p. 37).

36. 'The anatomical work of Galen was sufficiently adequate to fulfil the intellectual demands of of anatomists over fourteen centuries, during which the history of anatomy is one continuous line of exposition of the Galenic point of view'; L. Alberti López, *La Anatomía y los Anatomistas Españoles del Renacimiento* (Madrid: CSIC, 1948), pp. 42–3. Others write: 'Galen's work reaches its maximum heights in the sixteenth century when numerous Latin versions were printed throughout Europe … Despite some highly valid studies, we still do not possess a complete panorama of Galenism and Spanish medical humanism for the sixteenth century'; J. Riera Palmero, 'Andrés Laguna y el Galenismo Renacentista', in J. L. García Hourcade and J. M. Moreno Yuste (eds), *Andrés Laguna. Humanismo, Ciencia y Política en la Europa Renacentista* (Valladolid: Junta de Castilla y León, Consejería de Educación y Cultura, 2001), p. 166.

37. On the polemic between Spanish Renaissance medicine and the Arabic medical tradition, see L. García-Ballester, *Los Moriscos y la Medicina. Un capítulo de la medicina y la ciencia marginadas en la España del siglo XVI* (Barcelona: Labor, 1984), pp. 19–46. See also L. García-Ballester, 'The Circulation and Use of Medical Manuscripts in Arabic in Sixteenth-Century Spain', in J. Arrizabalaga, M. Cabré, L. Cifuentes and F. Salmón (eds), *Galen and Galenism: Theory and Medical Practice from Antiquity to the European Renaissance* (Aldershot and Burlington, VT: Ashgate, 2002), pp. 183–99.

38. '"Hippocratic" Galenism evident in the translations commented on by Mena, Cristóbal de Vega and Valles bears witness to the orientation of medical humanism in the second half of the sixteenth century. This made it possible, López Piñero has written, to convert Hippocratic texts into the model of medical knowledge and practice without questioning the authority of Galen'; L. S. Granjel, *La Medicina Española Renacentista* (Salamanca: Universidad de Salamanca, 1980), p. 32. 'Valles was, in reality, one of the main European figures in a tendency that emerged in the humanist current. This tendency, without questioning Galen and the authority of his system, transformed Hippocrates into the principal reference for medical knowledge and practice. Such a position, which we can call "Hippocratic" Galenism drew on Hippocratic texts through

the lens of humanism'; J. M. López Piñero, T. F. Glick, V. Navarro Brótons and E. Portela Marco, 'Valles, Francisco', in *Diccionario Histórico de la Ciencia Moderna en España*, 2 vols (Barcelona, Península, 1983), vol. 2, p. 392. 'Renaissance medicine returns to the texts of Hippocrates and Galen as primary sources … Hippocrates and Galen are the two most cited authors in [medical] compositions. The quotations are precise and may incorporate references to the works of textual critique to which [the original texts] were submitted', states M. J. Pérez Ibáñez, *El Humanismo Médico del Siglo XVI en la Universidad de Salamanca* (Valladolid: Universidad de Valladolid, 1997), pp. 114–15.

39.   B. Montaña de Monserrate, *Libro de la Anathomía del Hombre* (Valladolid: S. Martínez, 1551), fol. 61v, wrote that 'cabe notar que la mujer es diferente del varón, y por esta razón no pudo naturaleza echar fuera del vientre los miembros de la generación como el varón el cual por la fuerza de su calor pudo echarlos fuera' (it is necessary to note that woman is different from man and that for this reason nature cannot expell from inside the organs of generation as in man who, because of his heat, was able to expell them).

40.   J. Valverde de Amusco, *Historia de la Composición del Cuerpo Humano* (Rome: J. A. de Salamanca and A. Lafrery, 1556), fol. 65v, acknowledged that women's parts were inside the body to aid procreation. Valverde was a disciple of Realdo Colombo, whose master was Vesalio. Thomas Laqueur reproduces some of the laminates that appeared in the work by Valverde – which were renowned in his time – in order to show how dominant the one-sex model was (cf. Laqueur, *Making Sex*, pp. 76–7).

41.   J. Fragoso, *Erotemas Chirúrgicos* (Madrid: Sebastián Yáñez, 1570), fol. 26v, also acknowledged that Nature had made men's parts outside of their bodies as they were 'más calientes en la composición' (warmer in their composition). Later he argued, following Galen, that 'no hay más diferencia entre el hombre y la mujer cuanto a los miembros de la generación … sino que el hombre los tiene de fuera y la mujer dentro' (there is no difference between the organs of generation in men and women other than the fact that men have them outside and women inside); J. Fragoso, *Cirugía Universal* (1581; Madrid, 1627), p. 21. This work 'was highly appreciated and had fourteen Spanish editions and three in Italian in slightly more than a century' (López Piñero et al., 'Fragoso, Juan', in *Diccionario Histórico de la Ciencia Moderna en España*, vol. 1, p. 356).

42.   A. de León, *Libro Primero de Anatomía: recopilaciones y examen general* (Baeza: Juan Baptista de Montoya, 1590), fol. 28v: 'Háse de advertir, que las mujeres también tienen testículos y miembro viril, aunque ocultos y escondidos, y redondos como los de los hombres' (It is necessary to point out that women also possess testicles and a virile member, although they are hidden and covered and round like those of men).

43.   'Dicen los filósofos que la mujer no es otra cosa sino un varón imperfecto' (The philosophers say that woman is nothing but an imperfect man); B. Álvarez de Miravall, *Libro intitulado La Conservación de la Salud del Cuerpo y del Alma* (Medina del Campo: S. del Canto, 1597), fol. 286r. On the importance of this text, see L. S. Granjel, 'La Obra de Álvarez de Miraval', in *Médicos Españoles* (Salamanca: Universidad de Salamanca, 1967), pp. 93–116.

44.   J. Huarte de San Juan, *Examen de Ingenios para las Ciencias* (1575; Madrid: Editora Nacional, 1976), p. 315, argued, following Galen, that man was only different from women in that he possessed his genitalia outside his body. This was because 'si hacemos anatomía de una doncella, hallaremos que tiene dentro de sí dos testículos, dos vasos seminarios, y el útero con la mesma compostura que el miembro viril sin faltarle ninguna delineación' (if we undertake an anatomical examination of a lady, we will find within two testicles, two seminal tubes and a uterus with the same composition as a virile member with no structural difference). This work 'enjoyed several reprintings in few years – Pam-

plona, 1578; Bilbao, 1580; Valencia, 1580; Huesca, 1581, as well as translations into the principal European languages'; E. Torre, 'Introducción', in J. Huarte de San Juan, *Examen de Ingenios para las Ciencias* (Madrid: Editora Nacional, 1976), pp. 9–45, on p. 17.

45. Huarte de San Juan, *Examen de Ingenios*, p. 328: 'Cuál sea la correspondencia que han de tener el hombre y la mujer para que haya generación, dícelo Hipócrates de esta manera ... como si dijera: "si no se juntaren dos simientes en el útero de la mujer, la una caliente y la otra fría, o la una húmida y la otra seca, en igual grado de intensión, ninguna cosa se engendrará"' (Whatever correspondence man and woman must have in order to effect generation is as if Hippocrates were to state it thus: 'if two types of semen do not come together in the uterus of the woman, one hot and one cold, or one humid and one dry, in equal measure, nothing will be engendered'). In the same way, Álvarez de Miravall argues that the 'mujer incorrupta' (uncorrupted woman) 'no ha admitido mezcla de simientes' (has had no mixing of semen) (Álvarez de Miravall, *Libro intitulado La Conservación de la Salud del Cuerpo y del Alma*, fol. 471r). Here, a juridical consequence of Hippocratic doctrine is suggested. A woman who is pregnant and who accuses a man of rape is lying since women's pleasure is required for fertilization to take place; see P. Iglesias Aparicio, *Construcción del sexo y género desde la Antigüedad al siglo XIX* (Vigo: Universidad de Vigo, 2007), pp. 67–8, at http://webs.uvigo.es/pmayobre/textos/pilar_iglesias_aparicio/tesis_doctoral/cap2_construccion_de_sexo_y_genero_desde_%20la_edad_media.doc [accessed 14 April 2013].

46. L. Lobera de Ávila, *Libro de Regimiento de la Salud y de la esterilidad de los hombres y mujeres* (Valladolid: Sebastián Martínez, 1551), fol. 58r, writes: 'Según Hipócrates, el niño suele por la mayor parte engendrarse en el seno derecho de la madre y la hembra en la célula o seno izquierdo' (According to Hippocrates, a boy child is, for the most part, engendered in the right-hand cavity of the mother and the girl child in the left cell or cavity). Lobera de Ávila asserts that women who usually sleep on the right-hand side rarely give birth to girl babies (ibid., fol. 38v). Luis Lobera de Ávila (fl. 1530) was a doctor in the service of Carlos V (Alberti López, *La Anatomía*, pp. 61–7). Huarte de San Juan, for his part, (*Examen de Ingenios*, pp. 333–6), offers advice on how to conceive male children.

47. Lobera de Ávila, *Libro de Regimiento de la Salud*, fol. 33v. Álvarez de Miraval writes that women have an 'expultrix' facility with respect to semen, situated in the testicles, and an 'attractrix' impulse (absent in males) with respect to male semen once in the womb (Álvarez de Miravall, *Libro intitulado La Conservación de la Salud del Cuerpo y del Alma*, fols 280r–v).

48. 'la simiente de la mujer no tiene nunca aquella perfección de adherencia y viscosidad ... lo cual consta por experiencia pues que para la generación del cuerpo humano es tan necesaria la simiente del varón que no se puede engendrar sin ella' (the seed of woman never possesses that perfection of adherence and viscosity ... which is shown by experience since the male seed is so necessary that procreation is not possible without it) (Montaña de Monserrate, *Libro de la Anathomía del Hombre*, fols 65r–66v). Montaña opposes the Hippocratic division of the womb into cavities and believed that 'se halla en ella más que una cavidad muy pequeña' (there was only a very small cavity to be found there) (ibid., fol. 72r). According to historians Jacquart and Thomasset, this belief in the cavities of the uterus was derived from the idea of the female womb that certain animals possessed, in particular pigs. See D. Jacquart and C. Thomasset, *Sexualidad y Saber Médico en la Edad Media* (Barcelona: Labor, 1989), pp. 147–9.

49. Montaña de Monserrate defended this notion: the foetus was formed by the fermentation of the semen (mainly male but not excluding the female) with menstrual blood. With arterial blood the solid parts of the body were formed, and with the blood of the veins the

softer parts were formed. See his *Libro de la Anathomía del Hombre*, fols 61v–62r. This distinction goes back to Alî Ibn al-Abbas (Iglesias Aparicio, *Construcción del sexo*, p. 64).

50. Valverde de Amusco, *Historia de la Composición del Cuerpo Humano*, fol. 65r, also wrote that the male seed was the most important in the process, although he recognized that the female input was vital and no less fertile (ibid., fol. 65v).

51. 'Y, así, cuenta Hipócrates que los hombres principales de Escitia eran muy afeminados, mujeriles, mariosos, inclinados a hacer obras de mujeres como son barrer, fregar y amasar. Y, con esto, eran impotentes para engendrar; y si algún hijo varón les nacía, o salía eunuco o hermafrodita' (And, so, Hippocrates writes that the most important men among the Scythians were very effeminate, womanly, soft and inclined to perform women's tasks such as sweeping, cleaning and making bread. And, as a result of this, they were impotent and could not engender; if a boy child was born to them, he was either a eunuch or a hermaphrodite) (Huarte de San Juan, *Examen de Ingenios*, p. 336). On the Scythians, see also P. de Peramato, *Opera Medicinalia: De elementis, De humoribus, De temporamentis* (Sanlúcar de Barrameda: Fernando Díaz Imp., 1576), fol. 67.

52. The classical taxonomy, established in the Hellenistic period by Paul of Aegina, distinguished four types of hermaphrodite, three of which were masculine and one feminine (Marchetti, *L'invenzione della bisessualità*, p. 73). The Portuguese Pedro de Peramato (fl. 1575), educated at the University of Salamanca, classified hermaphrodites according to four types. Three were predominantly of the masculine type and one feminine; see Peramato, *Opera Medicinalia*, fol. 117r–v. On Peramato, see A. Chinchilla, *Anales históricos de la medicina en general y biográficos-bibliográficos de la española en particular: Historia de la Medicina Española*, 4 vols (New York and London: Johnson Reprint Corporation, 1967), vol. 2, pp. 77–81. This taxonomy can be compared with that established by Albucasis: there were two types of male hermaphrodite (one with the appearance of female genitalia with hair in the region of the perineum, and another with analogous genitalia but these were situated on the skin of the scrotum, between the testicles), and a third female type with smaller male genitalia situated at the height of the pubic hair. Albucasis suggests an Aristotelian explanation as he alludes to excessive growth. The difference with respect to Peramato is that Albucasis considers a variant of his second group (those that urinate through apparent female genitalia attached to the scrotum) those which Peramato identifies as a different class altogether. See Albucasis, *On Surgery and Instruments*, ed. M. S. Spink and G. L. Lewis (London: Wellcome Institute of the History of Medicine, 1973), pp. 454–5. Avicenna, for his part, adds to the four types elaborated by Paul of Aegina two new varieties: one is neutral (no genitalia) and the other is perfect (possesses both sets of complete genitalia) (Marchetti, *L'invenzione della bisessualità*, pp. 73–4); and he considers that hermaphroditism in many cases can be cured surgically (see Jacquart and Thomasset, *Sexualidad y Saber Médico*, p. 149).

53. J. Sánchez Valdés de la Plata, *Crónica y Historia General del Hombre* (Madrid: Miguel Martínez, 1598), fol. 130r, wrote: 'Mas que aya hombres de entrambas naturas y que los haya habido no es cosa de dexar de sabello ... Yo vi en Salamanca un hijo de un caballero, que tenía entrambos miembros de hombre y mujer' (But that there are and have been men with both natures is not out of the question ... I saw in Salamanca the son of a gentleman who had members of both a man and a woman) (the text was written between 1545 and 1550). The book by the La Mancha doctor Juan Sánchez Valdés de la Plata, as found by Elena Ronzón, is in part a plagiarized version of the often published *Silva de varia lección*, by Pedro Mexía, which appeared in 1540 (cf. E. Ronzón, 'El Médico Juan Sánchez Valdés de la Plata y su libro sobre el Hombre. Historia de una investigación', *El*

Basilisco, 24 (1998), pp. 63–84). However, the chapter by Sánchez Valdés de la Plata on sex changes does not appear in the work by Mexía.

54. On this opinion, see Marchetti, *L'invenzione della bisessualità*, p. 114. Luis Mercado, considered to be the 'Saint Thomas of Medicine', was responsible for the systematization in sixteenth-century Spain of traditional medieval medical doctrines (see López Piñero et al., 'Mercado, Luis', in *Diccionario Histórico de la Ciencia Moderna en España*, vol. 2, pp. 56–9). In Chapter 6 ('De monstroso conceptu') of the third book of *De Mulierum Affectionibus* hermaphroditism is not examined extensively. It is discussed as an example of monstrosity, together with other forms of the same, in order to show the diversity of this phenomenon. See L. Mercado, *De Mulierum Affectionibus* (Valladolid: Diego Fernández de Córdoba, 1579), p. 385.

55. Mercado, *De Mulierum Affectionibus*, pp. 387–94.

56. Ibid., p. 385.

57. Ibid., p. 235.

58. Andrés Laguna (1510–59), doctor to the court of Fernando el Católico and in the service of Cardinal Francisco de Bobadilla y Mendoza from 1545, cared for Popes Paulo III and Julio III. In a letter to the latter from 1551, Laguna wrote that he had seen with his own eyes men in Germany who lactated and menstruated. He did not consider them to be monstrous cases (cf. Pomata, 'Uomini Menstruanti', p. 57). On the career of Laguna, see L. S. Granjel, 'El Médico Andrés Laguna', in J. L. García Hourcade and J. M. Moreno Yuste (eds), *Andrés Laguna. Humanismo, Ciencia y Política en la Europa Renacentista* (Valladolid: Junta de Castilla y León, Consejería de Educación y Cultura, 2001), pp. 11–16.

59. Álvarez de Miravall, *Libro intitulado La Conservación de la Salud del Cuerpo y del Alma*, fol. 278v. Fragoso for his part, alluding to what was being rediscovered at the time as the clitoris, although not identifying it as such, wrote of certain women being in the possession of a kind of protuberance (citing Paul of Aegina, Avicenna and Albucasis as authorities) similar to the male organ. This 'suele crecer tanto, que las mujeres que la tienen se juntan con las otras como si fuesen hombres' (often grows to such dimensions that those women who possess it lie with other women as if they were men) (Fragoso, *Cirugía Universal*, p. 21). Further on he refers to certain women whose organ, identified as the nymphs or labia, is as large as a man's (ibid., p. 564). This comment serves to introduce the case of the women of Thessalonika who 'se juntaban y tenían poluciones' (lay together and experienced pollution), to the degree that one of them, after having had relations with her husband, was united carnally with another widowed woman who became pregnant.

60. These were individuals who in the womb would have been female for several months. They were transformed into male foetuses through an excess in heat and would retain, on birth, certain behaviours and characteristics of their first sex, with womanly mannerisms, a melodious voice and a tendency to commit the nefarious sin (Fragoso, *Cirugía Universal*, p. 163). Huarte de San Juan, in a work published six years before that of Fragoso, had said practically the same thing (see Huarte de San Juan, *Examen de Ingenios*, pp. 315–16). Fifteen years later, Andrés de León repeated the same idea on the subject of 'varones amugerados en hechos y palabras' (womanly men in their acts and their words), with a tendency to commit the nefarious sin (de León, *Libro Primero de Anatomía*, fol. 28v). The same concept, in reverse, was utilized by these three authors for 'mujeres varoniles y machorras' (manly and virile women), in the terms used by Andrés de León.

61. In these cases, the cause was seen as a matter of diabolical intervention, witchcraft or divine punishment. See Velasco, *Male Delivery*, pp. 1–27 and 45.

62. As we have stated, Andrés Laguna mentioned that he had seen numerous cases of lactating men. Fuentelapeña wrote that Bernardino Montaña de Montserrate had also seen cases of men who breastfed their children (Fuentelapeña, *El Ente Dilucidado*, p. 253). In a study published in 1880, Dr Ángel Pulido published an extensive study of such cases, drawing on examples from the sixteenth and seventeenth centuries (Benedicto, Rhodio, Whaleo, Cardano, Bartholino, Robert), but with no mention of Spanish doctors; A. Pulido, 'Lactancia Paterna', *Revista de Medicina y Cirugía Prácticas*, 7 (1880), pp. 13–22, on p. 15.

63. Fragoso, *Cirugía Universal*, pp. 162–3, devotes a chapter to the question. He admits the possibility of change in both directions. Huarte de San Juan, *Examen de Ingenios*, pp. 315–16, argues the same. Sánchez Valdés de la Plata, *Crónica y Historia General del Hombre*, fol. 17v, invokes Aristotelian teleology in order to reject episodes of supposed feminization. Peramato, *Opera Medicinalia*, fol. 117r, mentions two cases in Spain, seeing them not as sex changes in the true sense as described by Pliny, but as the expulsion of a sex that was hidden. These were what would later be described as 'hidden hermaphrodites'. Álvarez de Miravall, *Libro intitulado La Conservación de la Salud del Cuerpo y del Alma*, fol. 286r, also casts doubt on cases of feminization: 'Y así leemos que las hembras se muden en varones, y que muy raras veces se ha visto lo contrario, porque siempre la Naturaleza procede de lo más malo a lo mejor, y no al contrario' (And so we read about women who change into men and very rarely have we seen the contrary, because Nature always proceeds from the worst towards the best, and not the reverse). He cites all the Greco-Roman authorities on the question of masculinization (Hippocrates, Lucinio Murciano, Pliny, Aulo Gelio, Ovid, Tito Livio). De León, *Libro Primero de Anatomía*, fol. 28v, describes sex changes in both directions, but only in the womb. He concludes by praising Providence for the creation of men and women: 'Mas al fin fue gran providencia de Dios y Naturaleza, que hubiera varones y hembras, porque de otra fuese, Naturaleza cesara ... Que Dios y Naturaleza no hacen cosa en vano' (But, in the last analysis, it was a great providence from God and Nature that there were men and women because, otherwise, Nature would cease ... God and Nature do nothing in vain) (ibid., fol. 28v). Here the same kind of ambivalence as noted at the beginning of the chapter is present. God permits 'marvels' and at the same time guarantees the division between male and female in order for procreation to result.

64. See also Velasco, *Male Delivery*, pp. 93–4, and Marchetti, *L'invenzione della bisessualità*, pp. 129–31.

65. See del Río Parra, *Una era de monstruos*, pp. 97–100, and Velasco, *Male Delivery*, pp. 93–7.

66. Marchetti, *L'invenzione della bisessualità*, pp. 124–5.

67. Cf. Park, 'The Rediscovery of the Clitoris', pp. 179–83.

68. Stolberg, 'A Woman Down to her Bones'.

69. Laqueur, 'Sex in the Flesh', pp. 301–3.

70. J. López Piñero et al., 'García Carrero, Pedro', in *Diccionario Histórico de la Ciencia Moderna en España*, vol. 1, p. 374. See also L. S. Granjel, *La Medicina Española del siglo XVII* (Salamanca: Universidad de Salamanca, 1978), p. 24: 'the importance of the work of Pedro García Carrero for the evolution of medicine in the seventeenth century is confirmed by the continued reference to his writings throughout the century and the major influence of his teachings'.

71. P. García Carrero, *Disputationes Medicae super Libri Galeni de Loci Affectis et de Alii Morbis ab eo Relictis* (Alcalá de Henares: J. Sánchez Crespo, 1605), p. 1034.

72. Ibid., p. 1035.

73. Ibid., p. 1181.

74. Ibid.
75. Daston and Park, 'The Hermaphrodite and the Orders of Nature' (1995), p. 424.
76. In this period, as the *Tesoro de la Lengua Española* by Sebastián de Covarrubias shows, the two theories were current. The Aristotelian theory that identified the hermaphrodite as a type of monster and the Hippocratic theory that considered the hermaphrodite as an unusual but not monstrous product of generation were both present (Covarrubias, 'Andrógeno', in *Tesoro de la Lengua Castellana o Española*, p. 118).
77. García Carrero, *Disputationes Medicae*, p. 1181.
78. For example, ibid., pp. 1054 and 1115.
79. Marchetti, *L'invenzione della bisessualità*, pp. 145–8.
80. We have used the 1630 edition with the title *De Partu Naturali et Legitimo* (Cologne: Jacobi Stor). The importance of Carranza for medicine is shown by the fact that Anastasio Morejón provides an entry on him in Chinchilla, *Anales históricos de la medicina en general*, vol. 2, pp. 330–1.
81. R. a Castro Lusitani, *De universa muliebrium morborum medicina*, 3rd edn (Hamburg: Ex bibliopolio Frobbeniano, 1628), p. 144 (1st edn 1603).
82. Carranza, *De Partu Naturali et Legitimo*, p. 645.
83. On the translation of 'peccatum' as an error of Nature and not as a human sin in this context, see Daston and Park, *Wonders and the Order of Nature*, p. 201.
84. Carranza, *De Partu Naturali et Legitimo*, p. 647; this argued against the thought of Clement of Alexandria.
85. Ibid., pp. 645–6.
86. Bravo de Sobremonte was 'The most representative and prestigious figure of moderate Galenism in Spain in the mid-seventeenth century. Numerous elements of modern doctrines were accepted by him' (López Piñero et al., 'Bravo de Sobremonte Rodriíguez, Gaspar', in *Diccionario Histórico de la Ciencia Moderna en España*, vol. 1, p. 133). Cf. also Granjel, 'La Obra de Gaspar Bravo de Sobremonte', in *Médicos Españoles*, pp. 133–64, esp. p. 134.
87. G. Bravo de Sobremonte, 'Promptuarium XXIV', 'De sexus mutatione' and 'De hermaphroditis', in *Operum Medicinalium. Tomus Tertius* (Lyon: L. Arnaud, 1671), pp. 246–9 and 249–50; and G. Bravo de Sobremonte, 'Resolutio I. Utrum sexus transmutatio permitatur naturae', in *Operum Medicinalium. Tomus Quartus: Tres Disputationes Complectens* (Lyon: L. Arnaud, 1679), p. 198.
88. Bravo de Sobremonte, 'Resolutio I. Utrum sexus transmutatio permitatur naturae', p. 198. In the same promptuarium, published eight years earlier, those authorities that gave credence to cases of masculinization are mentioned. The reverse case scenario appears to be rejected from the start. Later, citing Luis Mercado and André du Laurens, among others, three arguments that refute such a possibility are outlined. Women transformed into men are in reality hermaphrodites whose male members emerge late. They may also be bearded women as a result of menstrual alterations or a kind of deformation of their parts (Bravo de Sobremonte, 'Promptuarium XXIV', 'De sexus mutatione', p. 248). In this text, only the Madrid and Úbeda cases are mentioned. Both are cited by Peramato, and another case of metamorphosis in Villaviciosa (Madrid) is discussed (ibid., p. 247).
89. 'Natura per se intendit hominis generationem, speciei conservationem, et media, seu instrumenta simpliciter in hunc finem necessaria: sed uterque sexus est omnino et simpliciter necessarius, vir ut generet in alio, mulier ut in se generet pro conservanda specie humana. Ergo uterque sexus est per se intentus a natura' (Bravo de Sobremonte, 'Resolutio I. Utrum sexus transmutatio permitatur naturae', p. 198). 'Deus enim per se formavit foeminam, non intendens facere monstra, sed quae necessaria erant per se ad universi

pulchritudinem, et hominis conservationem' (Bravo de Sobremonte, 'Promptuarium XXIV', 'De sexus mutatione', p. 247).

90. 'Sed solum accidentaliter marem et foeminam differre' (Bravo de Sobremonte, 'Resolutio I. Utrum sexus transmutatio permitatur naturae', p. 197). This thesis, derived from the work of du Laurens, has been interpreted by Laqueur as a softening of the distinction between men and women and as a result still firmly within the one-sex paradigm (Laqueur, 'Sex in the Flesh', p. 303). Rather, we understand that the point should be understood in the context of the discussion (especially in the case of Bravo de Sobremonte); what is being insisted upon is that women are not monsters and that they do not belong to another species.

91. Bravo de Sobremonte, 'Resolutio I. Utrum sexus transmutatio permitatur naturae', p. 197.

92. The comparison concludes: 'Quia alia est viris fabrica, alii viri testes, alii testes mulieris, alia mulieris, aliud viri scrotum et alius mulieris uterus; aliud priapus et virga viri, aliud mulieris vagina inanis et vacua semper, non ita penis' (ibid., p. 199).

93. Ibid.: 'praecipue respondemus ex comprobationibus Clasicorum Medicorum, quas adduximus, ex quibus manifeste pacet impossibilitas mutationum partium dicatarum generationi, varia illarum conformatio, impossibilis adaptatio ab uno sexo in alium et inmutatio ab internis partibus, in quibus sunt a natura aligatae, in partes externas'.

94. 'Secunda fallendi causa sunt hermaphroditae seu androgynes, qui cum utroque gaudeant sexu, ubi sunt illo pro viris et pro foeminis, et ideo a viris in foeminas aut econtra judicabantur mutari' (ibid., pp. 199–200).

95. Ibid., p. 200. 'Ad secundum dicimus, tum pro primo, quod in re tam controversa nulla fides danda est fabulosis Poetarum figmentis' (Bravo de Sobremonte, 'Promptuarium XXIV', 'De sexus mutatione', p. 249).

96. On the attitude of these authors to hermaphroditism, see Park, 'The Rediscovery of the Clitoris', pp. 178–84.

97. 'Praeter hos variae hermaphroditorum observantur, ac monstruosae conformationes … Haly Rhodoan vidit puerum sine vulva, cum virili miembro, et sine testibus, qui levi orificio reddebat lotium. Duplicem penem vidit Schenkius in alio, observatque in puella duplicem vulvam' (Bravo de Sobremonte, 'Promptuarium XXIV', 'De hermaphroditis', p. 249).

98. 'Ex iis androgynorum inchoatis conformationibus credendum est, quos aliqua membra virilia, quae intus latitantia errumpere non valuerunt, tractu temporis erumpant ad extra, vigore per aetates acquisito, et ideo quae prius apparentibus partibus: sed tamen androgyni semper debent haberi, non per instrumenta, vel partes generationi dicatas de novo apparentes' (ibid., p. 250).

99. Around 1632, doctors such as the court physician Juan de Quiñones and Gerónimo de la Huerta had no difficulty in affirming that male Jews menstruated periodically, just like women. See J. L. Beusterien, 'Jewish Male Menstruation in Seventeenth-Century Spain', *Bulletin of the History of Medicine*, 73:3 (1999), pp. 447–56. One way of according infamy to Jews was precisely by attributing to them the category of 'imperfect males', that is, men who are like women. A tract published in Barcelona in 1606 refers to the case of a Granada-born man who, as a result of an act of sorcery, gave birth to a foetus with the form of the devil. On this case, see P. Cordoba, 'L'homme enceint de Grenade. Contribution à un dossier d'histoire culturelle', *Mélanges de la Casa de Velázquez*, 23 (1987), pp. 307–30, and Salamanca Ballesteros, *Monstruos*, p. 311.

100. Granjel, *La Medicina Española del siglo XVII*, p. 149, refers to this work as 'the most important single contribution to the medical literature of the seventeenth century on the subject of congenital anomalies'.

101. J. Rivilla Bonet, *Desvíos de la Naturaleza o Tratado del Origen de los Monstruos* (Lima: Imprenta Real, 1695).

102. Ibid., fol. 8r.

103. Ibid., fol. 9v.

104. Ibid., fol. 14v.

105. Ibid., fols 35v–36r.

106. Sebastián de Covarrubias also contemplates this cause when he refers to 'andrógenos' or 'ermaphroditos' (Covarrubias, 'Andrógeno', in *Tesoro de la Lengua Castellana o Española*, p. 118).

107. Rivilla Bonet, *Desvíos de la Naturaleza*, fol. 37v.

108. The jurist Alonso Carranza severely rejects the hermetic and gnostic theories of the first man created by God as a hermaphrodite: 'Exterminentur ergo ista deliria et consimilia existimantium, Adamum Androgynum fuisse; ut qui non multum denient a fabuloso antiquo errore' (Carranza, *De Partu Naturali et Legitimo*, p. 645). On the beliefs of some early Christian sects on Adam and Eve as a 'parted androgyne', see P. Brown, *The Body and Society: Men, Women and Sexual Renunciation in Early Christianity* (New York: Columbia University Press, 1988), p. 268. Antonio de Fuentelapeña alludes to heresy by stating, 'si Adán no hubiera pecado, no hubiera división de sexos' (if Adam had not sinned, there would not have been a division of the sexes) (Fuentelapeña, *El Ente Dilucidado*, p. 164).

109. It was Jacques Derrida who first elaborated the idea of 'resignification' as the reiteration of deconstructive effects. The idea has been most used by queer studies and by Judith Butler. On the need of 'resignifying' without losing the terms 'gender' and 'sex' while also realizing that the terms ('race' included) are part of oppressive regimes, see Butler, *Bodies that Matter*, p. 123: 'precisely because such terms have been produced and constrained within such regimes, they ought to be repeated in directions that reverse and displace their originating aims'.

110. M. Delcourt, *Hermafrodita. Mitos y Ritos de la Sexualidad en la Antigüedad Clásica* (Barcelona: Seix Barral, 1970), pp. 67–76. On monsters as negative auguries, see Salamanca Ballesteros, *Monstruos*, pp. 171–200.

111. Marchetti, *L'invenzione della bisessualità*, pp. 5 and 187–8.

112. The recopilation undertaken by Gaspar Bauhinus (1614) makes reference to some theologians who understood that coitus with a demonic being could be a possible cause of hermaphroditism (ibid., p. 15).

113. J. Boswell, *Chrisitianity, Social Tolerance and Homosexuality: Gay People in Western Europe from the Beginning of the Christian Era to the Fourteenth Century* (Chicago, IL and London: Chicago University Press, 1980), pp. 356–7.

114. Cf. D. Wilson, *Signs and Portents: Monstruous Births from the Middle Ages to the Enlightenment* (London and New York: Routledge, 1993), pp. 65–77, Daston and Park, *Wonders and the Order of Nature*, p. 183, and Salamanca Ballesteros, *Monstruos*, pp. 171–200.

115. Gutiérrez de Torres, *El Sumario de las Maravillosas*, fol. a 5.

116. Ibid., fol. b 6.

117. Ibid., fol. f 1.

118. Daston and Park, *Wonders and the Order of Nature*, pp. 177–80; Salamanca Ballesteros, *Monstruos*, pp. 192–4; del Río Parra, *Una era de monstruos*, pp. 64, 88.

119. M. Alemán, *Guzmán de Alfarache*, in A. Valbuena y Prat (ed.), *La Novela Picaresca Española* (Madrid: Aguilar, 1968), pp. 159–637, on p. 246. Rivilla Bonet, *Desvíos de la Naturaleza*, fol. 37r, also mentions the monster of Ravenna and records that 'otras mon-

struosidades acaecen en predictio y aviso de futura venganza y males grandes, como han precedido en varias ocasiones a sangrientas guerras y herejías' (other monstrosities occur as predictions and warnings of future avengements and great evils, and have predicted bloodly wars and acts of heresy on various occasions) (fol. 36v). The deciphering of the significance of monstrous births, as well as the interpretation of certain dreams, could be valuable in terms of political predictions, as in the example of the praise given to Mateo Vázquez de Leca, the secretary of Felipe II, by Calvete de Estrella. In this text a monster born in Ledesma is referred to, and it is seen as a bad sign for the monarchy: 'Muéstranos Urania, ornato del firmamento e ilustre hija de Júpiter, qué anuncian los fieros monstruos y qué crueles peligros y desastres nos presagian a los angustiados mortales' (Reveal to us, o Urania, adnornment of the firmament and illustrious daughter of Jupiter, what the savage monsters have announced and what cruel dangers and disasters are presaged for we anguished mortals); J. C. Calvete de Estrella, 'Elogio del Secretario Real Don Mateo Vázquez de Leca' (orig. *c.* 1585), cited in M. A. Díaz Gito, 'El Poema "Corsica" de J. Cristóbal Calvete de Estrella (y otros dos poemas latinos)' (unpublished doctoral thesis, Universidad de Cádiz, 1990), p. 21.

120. 'la criatura que nació era niño y niña, con dos naturalezas, la de niña en la parte común y la de niño en mitad de la frente, cosa la más espantosa que los nacidos han visto' (the child that was born was a boy and a girl, with two natures, that of the girl in its usual place and that of the boy in the middle of the forehead, the most horrendous thing that mortals have seen) (*Relación Verdadera y Caso Prodigioso y Raro, que ha sucedido en esta Corte el día catorce de mayo de este año de 1688*, in Ettinghausen, *Noticias del Siglo XVII*, n.p.).

121. The deformity would serve 'de exemplar a todos los Católicos Christianos, por si en su generación huvo algún excesso vicioso, que suele el Cielo castigar en los hijos travesuras y desacatos de los padres' (as an example to all Catholic Christians, for if there were any vicious excesses in the act of generation, Heaven will usually visit the misdemeanours and carelessness of the parents on their children) (ibid., n.p.).

122. This reading proposed by Morel d'Arleux, 'Las "Relaciones de Hermafroditas"', p. 270, is a little forced if we go back to the origianl text. Here, it is only said that in the court of Carlos II news of these cases from 'Reynos muy distantes' (far-off kingdoms) is extensive, but none comes from Spain. Obviously, the consumption of this kind of news may well have been quite common, given the precarious nature of the Hapsburg monarchy, but to extrapolate a political sign from it appears to be excessive.

123. Gilbert, *Early Modern Hermaphrodites*, pp. 65–6 and 104–6, and Long, *Hermaphrodites in Renaissance Europe*, pp. 189–213 on cases in England and France, respectively.

124. Fernando R. de la Flor has analysed these sources and indicates that around 1555, Carlos V had sought from Sánchez Coello two portraits of bearded women, now lost. To these we can add the also lost account by Diego Valentín Díaz of 1660 and, especially, those of José de Ribera ('Magdalena de los Abruzos', 1631) and Juan Sánchez Cotán ('Brígida del Río', 1590). Brígida del Río (mentioned, as we have said, by Bravo de Sobremonte in 1679; de la Flor points to references from Mateo Alemán in 1599, Covarrubias in 1611 and Jerónimo de Alcalá in 1624) was popular in her time. The portrait by Sánchez Cotán was made into numerous copies, and other painters also depicted her. Magdalena was exhibited in the court of the Viceroy of Naples (see de la Flor, 'La "puella pilosa"', pp. 269–305). Many copies of the portrait by Sánchez Cotán circulated at the time, as well as versions by other artists. Covarrubias (*Tesoro de la Lengua Castellana o Española*, p. 164) comments on an emblem (no. 64) which represented this bearded woman and the negative associations that accompanied her. The myth of Hermaphrodito and Salmacis

is also mentioned in this context, and there is a suggestion that amorous passion effemin-
izes the male. The lines that accompany the drawing convey the same idea: 'Me tienen
por siniestro y mal agüero / Advierta cada cual que me ha mirado / que es otro yo / si
vive afeminado' (They have taken me as sinister and as a bad omen / Warn anyone who
has seen me / that it is another me / if in effeminate form). The same asssociations can
be seen in P. J. Smith, *The Body Hispanic: Gender and Sexuality in Spanish and Spanish
American Literature* (Oxford: Oxford University Press, 1989), p. 9.

125. De la Flor, 'La "puella pilosa"', pp. 286–8.

126. Alcalá Galán, 'El andrógino', pp. 116–17.

127. Federico Garza has presented this same argument against the interpretations of McIn-
tosh, Bray and Trumbach. See F. Garza Carvajal, *Quemando Mariposas. Sodomía e
Imperio en Andalucía y México, siglos XVI–XVII* (Barcelona: Laertes, 2002), pp. 96–7. As
Alcalá Galán, 'El andrógino', p. 111, and Martínez, 'Mari(c)ones, travestis y embrujados',
p. 27, have indicated, all sodomites were not associated with effeminacy, a description
often applied to the 'passive' partner, and all forms of effeminacy did not necessarily
include all sodomites. For this reason, we believe that the work by Velasco, *Male Delivery*, pp. 112–19, tends to link sodomy with effeminacy too mechanically, and although
she admits some exceptions to the rule, sodomy becomes too readily and anachronisti-
cally associated with homosexuality.

128. This rather utopian understanding has perhaps been encouraged by a superficial reading
of texts like that of Ambroise Paré. Paré may appear to suggest that any hermaphrodite
could choose their sex: 'ancient and modern laws have obliged and still oblige [hermaph-
rodites] to choose which sex organs they wish to use, and they are forbidden on pain
of death to use any but those they will have chosen, on account of the misfortunes that
could result from such'; A. Paré, *On Monsters and Marvels*, trans. J. L. Pallister (1575;
Chicago, IL: Chicago University Press, 1982), p. 27.

129. Marchetti, *L'invenzione della bisessualità*, p. 69. This is the best and most complete study
to date on the juridical controversies surrounding the question. This author is drawn
upon extensively in the ensuing section of this chapter.

130. Ibid., p. 72.

131. 'Hermafroditus en latín, tanto quiere decir en romance como aquél que ha natura de varón
e de mujer. E este atal, dezimos, que si tira más a natura de mujer, que de varón, non puede
ser testigo en testamento, nin en todas las otras mandas que ome fiziesse. Mas si se acostase
más a natura de varón, entonces bien pueda ser testigo en testamento en todas las otras
mandas que ome fiziesse' (Hermafroditus is said in Latin; in Romance the same is meant
as he who has the nature of a man and a woman. And this same person, we say, if he tends
more towards the nature of woman than of man, cannot be a witness in a testament or in
all other legal accords that a man may agree to); Alfonso X El Sabio, *Las Siete Partidas ...
con las variantes de más interés y con la glosa del Licenciado Gregorio López*, ed. I. Samponts,
vol. 4, Partida VI, Tit. 1, Ley 10 (Barcelona: Imprenta de Antonio Bergnés, 1844), p. 409.

132. Marchetti, *L'invenzione della bisessualità*, pp. 238–9.

133. Ibid., p. 137.

134. Ibid., pp. 233–4. The Franciscan E. de Villalobos, *Manual de Confesores* (Salamanca:
Imp. Diego de Cossío, 1628), pp. 162–3, put it like this: 'El hermafrodita, si prevalece
en un sexo, puede contraer matrimonio conforme a el, y si es igual en ambos sexos, puede
usar del que quisiere, y casarse conforme a el. Y hase de obligar con juramento de non
usar del otro sexo' (If the hermaphrodite prevails in one sex, he can contract matrimony

accordingly and if he is equal in the two sexes, he can use the one he wishes and marry accordingly. He must swear by oath that he will not use the other sex).

135. Marchetti, *L'invenzione della bisessualità*, pp. 245–6.

136. Ibid., p. 245.

137. Ibid., pp. 157–9. Torquemada, *Jardín de Flores Curiosas*, p. 116, mentions two cases of hermaphrodites who committed 'perjury' in Burgos and Seville. They went back on the sex they had chosen and were burned alive for sodomy: 'LUIS: y así a lo que he oído, en Burgos dieron a escoger a una que usase de la natura que quisiese y no de la otra, so pena de muerte; y ella escogió la de mujer, y después se averiguó usar secretamente la de hombre y hacer grandes maleficios debajo de esta cautela, y fue quemada por ello. ANT.: También se dice que en Sevilla quemaron a otra por lo mismo' (LUIS: as to what I have heard: in Burgos they allowed a woman to choose her sex and to use that nature and not her other nature under the pain of death. And she chose to be a woman and later it was discovered that she used secretly that of a man and that she performed bad spells under this guise, and she was burnt for it. ANT.: They also say that in Seville they burnt another woman for the same thing). All the more surprising and difficult legally was the marriage of two hermaphrodites in Valencia in 1662; the Valencian jurist Matheu i Sanz mentioned the case in his *Tractatus de Re criminali* (1686) in the section on hermaphrodites (as discussed in F. Tomás y Valiente, 'El Crimen y Pecado contra Natura', in F. Tomás y Valiente et al., *Sexo Barroco y otras transgresiones modernas* (Madrid: Alianza Universidad, 1990), pp. 33–55, on pp. 54–5).

138. T. Sánchez, *De sancto matrimonii sacramento disputationum*, vol. 3 (1607; Avenione: Hyeronimum Duperier, 1689), p. 381.

139. Marchetti, *L'invenzione della bisessualità*, p. 272. Carranza, *De Partu Naturali et Legitimo*, p. 646, points out that the 'Ancients' punished hermaphrodites with death; Catholics, however, point to eternal damnation.

140. Marchetti, *L'invenzione della bisessualità*, pp. 320–1.

141. On the concept of an 'act of the State' and its consequences in the social world, see P. Bourdieu, *Sur l'État. Cours au Collège de France 1989–1992* (Paris: Raisons d'Agir, 2012), pp. 26–30.

142. Marchetti, *L'invenzione della bisessualità*, p. 336. In the case of the hermaphrodite Thomas/Thomasine who was tried in colonial Virginia in 1629, the court, instead of pinpointing the 'predominant sex', determined that the two sexes were present. In this case, mixed clothing was prohibited as was all sexual activity on the basis that sodomy could have been committed. See Gilbert, *Early Modern Hermaphrodites*, pp. 46–8, and E. Reis, *Bodies in Doubt: An American History of Intersex* (Baltimore, MD: Johns Hopkins University Press, 2009), pp. 12–14.

143. Marchetti, *L'invenzione della bisessualità*, pp. 66–7.

144. See M. R. McVaugh, *Medicine before the Plague: Practitioners and their Patients in the Crown of Aragon, 1283–1345* (Cambridge: Cambridge University Press, 1993).

145. Marchetti, *L'invenzione della bisessualità*, p. 98.

146. This process in modern Spain has been discussed in T. Ortiz, 'From Hegemony to Subordination: Midwives in Early Modern Spain', in H. Marland (ed.), *The Art of Midwifery: Early Modern Midwives in Europe* (London: Routledge, 1994), pp. 95–114, and Velasco, *Male Delivery*, pp. 56–79.

147. Marchetti, *L'invenzione della bisessualità*, p. 165.

148. Ibid., pp. 119–20.

149. Ibid., p. 233.

150. The first to point to this highly original proposal by Matheu i Sanz was V. Marchetti, 'Propositions de règlement juridique d'une troisième sexualité: Lorenzo Matheu y Sanz et les hermaphrodites', in J. Poumarede and J. P. Royer (eds), *Droit, histoire et sexualité* (Lille: Université de Lille et Université de Toulouse, 1987), pp. 131–43.

151. F. Tomás y Valiente, 'Teoría y práctica de la tortura judicial en las obras de Lorenzo Matheu i Sanz (1618–1680)', in *La tortura en España* (Barcelona: Ariel, 1994), pp. 37–91, on pp. 37–47.

152. Marchetti, *L'invenzione della bisessualità*, p. 217.

153. On this case, see ibid., pp. 242–4. The whole controversy came as a result of the afore-mentioned scandal of the marriage of the Valencian hermaphrodites in 1662. See Tomás y Valiente, 'El Crimen y Pecado contra Natura', p. 54.

154. Marchetti, *L'invenzione della bisessualità*, p. 223.

155. Ibid., pp. 245, 294.

156. Ibid., p. 221.

157. Ibid., pp. 12, 269.

158. Ibid., p. 12.

159. Ibid., p. 294. Fuentelapeña, *El Ente Dilucidado*, pp. 181–6 and 187–90, provides a detailed discussion of these two alternatives in two of his sections on 'doubts': 'Si el hermafrodita, en quien prevalecen con igualdad y perfectamente ambos sexos, podrá a un mismo tiempo casarse con dos, esto es, con un hombre y una muger' (If the hermaphrodite in whom both sexes equally and perfectly prevail could at the same time marry two persons, that is, a man anad a woman), and 'Si ... sucesivamente podrá casar con diversos sexos dicho' (If ... the said hermaphrodite could marry successively with both sexes). Fuentelapeña, in ibid., p. 187, a work that is contemporaneous to that of Matheu, reminds the reader that the election must take place before a bishop or ecclesiastical judge.

160. On the importance of neo-Platonism in early Spanish modernity, see J. L. Villacañas, 'El cosmos intelectual de Villalobos. Sobre el carácter de la primera modernidad hispana' (2012), at http://saavedrafajardo.um.es/WEB/archivos/NOTAS/RES0119.pdf [accessed 3 May 2013], pp. 1–38, on p. 8.

161. Marchetti, *L'invenzione della bisessualità*, p. 224.

162. Pérez de Moya, *Philosophía Secreta*, fols 249r–251r.

163. Marchetti, *L'invenzione della bisessualità*, pp. 225–6.

164. Ibid., p. 224.

165. E. W. Cochrane, *Historians and Historiography in the Italian Renaissance* (Chicago, IL: University of Chicago Press, 1981), pp. 445–78; S. Bertelli, *Rebeldes, libertinos y orto-doxos en el Barroco* (Barcelona: Península, 1984), pp. 65–72; and B. Barret-Kriegel, *La défaite de l'érudition* (Paris: PUF, 1988), pp. 150–75.

166. Marchetti, *L'invenzione della bisessualità*, pp. 246–50.

167. Pérez de Moya, *Philosophía Secreta*, fol. 222v, noted that the moon was male and female simultaneously because sometimes active virtue operated, while at other times passive virtue held sway.

168. Marchetti, *L'invenzione della bisessualità*, pp. 253–5.

169. Ibid., p. 269.

170. The use of 'bisexual' by us here simply means that the subject could employ both sets of genitalia. The notion of 'bisexuality' as such, which in our view Marchetti, *L'invenzione della bisessualità*, utilizes with insufficient historical caution, was a much later invention and refers to the psychological type, the 'personality' of he or she who desires both sexes.

See S. Angelides, *A History of Bisexuality* (Chicago, IL: University of Chicago Press, 2001), pp. 23–48.

171. Tomás y Valiente, 'El Crimen y Pecado contra Natura', p. 54, and Marchetti, *L'invenzione della bisessualità*, pp. 272–3.

172. Gilbert, *Early Modern Hermaphrodites*, pp. 12–19, 81, argues that the idea of the androgyne, although present in English Renaissance thought in its neo-Platonic, Hermetic and alchemical expressions, declined towards the end of the seventeenth century. Long, *Hermaphrodites in Renaissance Europe*, pp. 109–62, developing French and German sources, examines how this same ideal remained present through the seventeenth century.

173. A clear example is showed by Pérez de Moya, *Philosophía Secreta*, fols 395v–196v, as del Río Parra, *Una era de monstruos*, p. 96, indicates. Caenis, daughter of Eletheo, was raped by Neptune, and the gods, to compensate for this act, offered her her greatest desire. Caenis asked 'que la tornase en forma de varón porque otra vez no padeciese semejante deshonra' (to be turned into a male form so that she would not suffer a similar dishonour once again) and became Caeneus. On the possible classical antecedents of the hagiographic discourse on saints who became men, see de la Flor, 'La "puella pilosa"', p. 298.

174. To these examples we should add another hagiographical repertoire, that of 'cross-dressed' saints, who lived as men in order to escape from the family or marital authority and to take up a life of perfection. This group includes Saints Eugenia, Anastasia, Margarita and Pelagia, whose sixth- and seventh-century lives are described in the Golden Legend (cf. S. Steinberg, *La Confusion des Sexes. Le travestissement de la Renaissance à la Révolution* (Paris: Fayard, 2001), p. 67; Greenblatt, *Shakespearean Negotiations*, p. 184). The behaviour of Saint Margarita seems to have inspired Joan of Arc (Steinberg, *La Confusion des Sexes*, p. 68).

175. Wilgefortis was probably born in Balcagia, now Bayona, in Pontevedra around AD 119. She was the daughter of Lucio Castelo Severo, Roman governor of Gallaecia and Lusitania (del Río Parra, *Una era de monstruos*, p. 96). Saint Wilgefortis was very popular in Spain in the fifteenth and sixteenth centuries, as the extensive iconography showing her martyrdom suggests. One of these representations can be found in the Museo Diocesano of la Seu d'Urgell. On Wilgefortis, see I. E. Friesen, *The Female Crucifix: Images of St. Wilgefortis since the Middle Ages* (Waterloo: Wilfrid Laurier University Press, 2001). Cf. also de la Flor, 'La "puella pilosa"', pp. 298–9. Strangely enough, of Saint Paula there remain no iconographies from the same centuries. The case of Santa Eulalia of Mérida is not discussed here (see Friesen, *The Female Crucifix*, p. 32).

176. 'Santa Paula, natural de Ávila, por librarse del furor de un cavallero, que dessatinadamente la amava, pidió a Dios la deformasse, y al punto la salieron barbas. En semejante trance Santa Liberata o Vilgefortis, hija del rey de Portugal impetró la misma disimulación, después fue crucificada por Christo' (Nieremberg, *Curiosa y Oculta Filosofía*, p. 55). On this tradition, see C. Vega, 'Figuras de la exclusión femenina: ermitañas, monjas visionarias, barbudas, mujeres pilosas' (unpublished paper given at the I Curso del Aula de Salamanca, May 1990, Universidad de Salamanca), and Cátedra, 'Sobre la ambigüedad'. For the case of Saint Gala, see de la Flor, 'La "puella pilosa"', p. 287.

177. Cátedra, 'Sobre la ambigüedad', p. 137.

178. See M. A. Valencia García, *Simbólica femenina y producción de contextos culturales. El caso de la Santa Barbada* (Ávila: Institución Gran Duque de Alba, 2004), for the most complete analysis. Strangely, there are no icons of Saint Paula remaining from the sixteenth to the eighteenth century; the sculpture that is used for the processions of 20 February is a contemporary piece.

179. Cátedra, 'Sobre la ambigüedad', p. 136.

180. Ibid., p. 133. The cult of Saint Paula was reactivated in the post-civil war period and she was venerated for her capacity for miraculous cures.

181. We are grateful to Chema Fraile, an expert in traditional Spanish culture, for bringing this case to our attention.

182. M. Comte Masía (ed.), *Casamiento entre dos damas* (1849; Madrid: Imp. de D. José Mª Marés, 2000), pp. 213–14. We would like to thank Chema Fraile for providing a copy of this text.

183. This subject has been studied by, among others, M. Romera Navarro, 'Las disfrazadas de varón en la comedia', *Hispanic Review*, 2:4 (1934), pp. 269–86; J. Homero Arjona, 'El disfraz varonil en Lope de Vega', *Bulletin Hispanique*, 39 (1937), pp. 120–45; Bravo-Villasante, *La Mujer Vestida de Hombre*; F. Castro Pires de Lima, *A Mulher vestida de Homem (Contribuição para o estudo do romance 'A Donzela que vai à Guerra')* (Coimbra: Fundação Nacional para a Alegria no Trabalho-Gabinete de Etnografia, 1958); B. B. Ashcomb, 'Concerning "la mujer en hábito de hombre" in the Comedia', *Hispanic Review*, 28:1 (1960), pp. 43–62; McKendrick, *Woman and Society*; K. Inamoto, 'La mujer vestida de hombre en el teatro de Cervantes', *Cervantes: Bulletin of the Cervantes Society of America*, 12:2 (1992), pp. 137–43; J. Sanz Hermida, 'Aspectos fisiológicos de la dueña dolorida: la metamorfosis de la mujer en hombre', in *Actas del Tercer Coloquio Internacional de la Asociación de Cervantistas* (Barcelona: Anthropos, 1993), pp. 463–72; B. Fuchs, 'Border Crossings: Transvestism and Passing in "Don Quijote"', *Cervantes: Bulletin of the Cervantes Society of America*, 16:2 (1996), pp. 4–28; R. A. Escalonilla López, 'Función del Travestismo en las Comedias de Don Pedro Calderón de la Barca' (unpublished doctoral thesis, Universidad Politécnica de Madrid, 1998); J. Cartagena-Calderón, '"Él es tan rara persona". Sobre cortesanos, lindos, sodomitas y otras masculinidades nefandas en la España de la temprana Edad Moderna', in M. J. Delgado and A. Saint-Saëns (eds), *Lesbianism and Homosexuality in Early Modern Spain* (New Orleans, LA: University Press of the South, 2000), pp. 139–76; P. Restrepo-Gautier, 'Afeminados, hechizados y hombres vestidos de mujer: la inversión sexual en algunos entremeses de los Siglos de Oro', in M. J. Delgado and A. Saint-Saëns (eds), *Lesbianism and Homosexuality in Early Modern Spain* (New Orleans, LA: University Press of the South, 2000), pp. 199–215; Velasco, 'Marimachos, hombrunas, barbados'; Velasco, *Male Delivery*; Alcalá Galán, 'El andrógino'; Martínez, 'Mari(c)ones, travestis y embrujados'.

184. On the distinction between forms of transvestism that reaffirm heterosexual culture and those that subvert it, see Butler, *Bodies that Matter*, pp. 124–37.

185. Cátedra, 'Sobre la ambigüedad', p. 136.

186. F. de Eiximenis, *Carro de las Donas, trata de la vida y muerte del hombre cristiano* (Valladolid: Juan de Villaquirán, 1542), fol. 29, cited in Morel d'Arleux, 'Las "Relaciones de Hermafroditas"', p. 271. See also Zamora Calvo, '*In virum mutata est*', p. 436. On the relation between illness and sins of luxury in Eiximenis, see M. Solomon, 'Fictions of Infection: Diseasing the Sexual Other in Francesc Eiximenis's *Lo Llibre de les dones*', in J. Blackmore and G. S. Hutcheson (eds), *Queer Iberia: Sexualities, Cultures and Crossings from the Middle Ages to the Renaissance* (Durham, NC: Duke University Press, 1999), pp. 277–90.

187. J. Eslava Galán, 'Introducción General', in *Cinco Tratados Españoles de Alquimia* (Madrid: Tecnos, 1987), pp. 13–47.

188. Authors such as Mircea Eliade and Jean Libis, by following (erroneously, in our view) Jungian theory on archetypes, find the androgyne in these primitive scenarios. See M. Eliade, *Mefistófeles y el Andrógino* (Barcelona: Labor, 1984), and J. Libis, *El Mito del Andrógino* (Madrid: Siruela, 2001). A critique of Jungian interpretations, which essentialize symbols without placing them in their specific contexts of domination, is F.

Vázquez García, 'Androginia y Pensamiento Esencial', *Culturas. Suplemento cultural del Diario de Sevilla*, 5 July 2001, p. 6. F. A. Yates, *The Rosicrucian Enlightenment* (London and New York: Routledge, 2002), p. 91, comments on *The Chemical Wedding*, supposedly written by Christian Rosencreutz, originally published in 1616, as 'an alchemical fantasia, using the fundamental image of elemental fusion, the marriage, the uniting of the *sponsus* and the *sponsa*' (male and female principles).

189. This alchemical symbol of the androgyne transcends strictly alchemical texts. See, for example, the illustrations that accompany the *relación* of the sexual metamorphosis of the nun Úbeda, published in 1617 (Morel d'Arleux, 'Las "Relaciones de Hermafroditas"', pp. 265–6). On the significance of the androgyne in the comedies *El Aquiles* and *La Dama del Olivar* by Tirso de Molina, see A. K. G. Paterson, 'Tirso de Molina and the Androgyne: "El Aquiles" y "La Dama del Olivar"', *Bulletin of Hispanic Studies*, 70:1 (1993), pp. 105–13. In *El Aquiles* a man dressed as a woman appears. This is unusual in the context of the Golden Age; the reverse is far more common. Aquiles, imitating the masking of other gods and demi-gods, is disguised as a woman in order to find his beloved Deidamia.

190. Eliade, *Mefistófeles y el Andrógino*, pp. 129–36; Libis, *El Mito del Andrógino*, pp. 58–64; Eslava Galán, 'Introducción General', pp. 36–7.

191. W. A. Meeks, 'The Image of the Androgyne: Some Uses of a Symbol in Earliest Christianity', *History of Religions*, 13:3 (1974), pp. 165–208.

192. Eliade, *Mefistófeles y el Andrógino*, pp. 131–2.

193. Libis, *El Mito del Andrógino*, pp. 104–5. Pedro Sánchez de Viana, drawing on the neo-Platonists León Hebreo and Marsilio Ficino, alludes to the androgynous condition of Adam, only to reject it immediately: 'Andrógino: Dios creó al primer hombre a su imagen, macho y hembra y púsole a ambos el nombre de Adam ... en el Paraíso Terrenal, Adam poseía las dos partes genitales, una delante y otra detrás, con lo qual esas dos partes no se tocaban ni podían unirse en ayuntamiento carnal, pero Sto. Tomás piensa que esto es falso y fueron los griegos, como Platón, los que volvieron a sacar esta fábula' (Androgyne: God created the first man in his image, male and female, and gave them both the name Adam ... in Earthly Paradise Adam possessed the two sets of genitalia, one in front and one behind, and these two parts neither touched one another nor could they be united in carnal intercourse. But Saint Thomas believes this is false and it was the Greeks, such as Plato, who revived this fable); P. Sánchez de Viana, *Anotaciones sobre los quinze libros de las Transformaciones, de Ovidio* (Valladolid: Diego Fernández de Córdoba, 1589), p. 3, cited in Morel d'Arleux, 'Las "Relaciones de Hermafroditas"', p. 266.

194. Eliade, *Mefistófeles y el Andrógino*, pp. 129–30, and Libis, *El Mito del Andrógino*, pp. 83–4.

195. On Adam as androgyne in the thought of Paracelsus, see A. Koyré, *Mystiques, spirituels, alchimistes du XVIᵉ siècle allemand* (Paris: Gallimard, 1971), p. 107.

196. On the Valencian Luis de Centellas (or Luis Centelles) (fl. 1552), see J. García Font, *Historia de la Alquimia en España* (Madrid: Editora Nacional, 1976), pp. 208–12, and J. Eslava Galán, 'La Alquimia Española del Siglo de Oro', in *Cinco Tratados Españoles de Alquimia* (Madrid: Tecnos, 1987), pp. 114–16.

197. L. de Centellas, *Coplas de Don Luis de Centellas sobre la Piedra Filosofal* (c. 1552), in *Cinco Tratados Españoles de Alquimia* (Madrid: Tecnos, 1987), p. 117. This 'lady' would be the feminine principle, associated with mercury, the generating force. Medieval Spanish alchemy would assign mercury an all-important role. However, in the alchemy of the Renaissance this hegemony would be transferred to sulphur, the masculine principle and active power of the solar essence (Eslava Galán, 'Introducción General', p. 31).

198. De Centellas, *Coplas*, pp. 117–18.

2 Sexual Transgression and Hermaphroditism: The 'New World' and Imperial Subjectivity

1. Apart from those cases mentioned in the following note, the remainder include (the authors or sources that mention them are in parentheses): Cordoba (Peramato, Fragoso, Bravo de Sobremonte), Santo Domingo del Real (Peramato, Fragoso, Pérez de Moya, Gómez de Huerta, Fuentelapeña), Peñaranda de Bracamonte (Mateo Alemán, Sebastián de Covarrubias, Jerónimo de Alcalá, Bravo de Sobremonte), Alcalá de Henares (the secular) (Nieremberg, Bravo de Sobremonte, Fuentelapeña), Alcalá de Henares (the religious) (Nieremberg, Fuentelapeña), Piedrahita (Ávila) (Sánchez Valdés de la Plata), Salamanca (Sánchez Valdés de la Plata), Benavente (Antonio de Torquemada, Juan de la Cerda, Martín del Río), Huesca (Martín del Río), Valdaracete (Madrid) (*Relaciones Topográficas*), Madrid (monstrous child, *Relación de suceso*), Burgos (Antonio de Torquemada), Seville (Antonio de Torquemada), Valencia (two hermaphrodites, Matheu i Sanz), Alhama (Granada) (Gerónimo de Huerta, Fuentelapeña), San Sebastián (Catalina de Erauso, Tascardo, Pérez de Montalbán), Toledo (Francisco Hernández), Zafra, Badajoz (the case of the priest, Juan Díaz Donoso, tried in 1634), Madrid (a hermaphrodite exhibited in Balverde Street as witnessed by Sánchez Tortolés in 1668) and Valencia (the garment maker referred to by Matheu i Sanz, tried in 1640).

2. The case that is thought to be somewhat dubious is that of the two nuns who supposedly changed sex, in Madrid probably in the mid-sixteenth century, and in Úbeda in 1617 (Fuentelapeña, *El Ente Dilucidado*, p. 245), respectively. The first of these, María Muñoz from the convent of Santo Domingo del Real (Madrid), who became a man, was ordained as a priest and took the name 'Rodrigo Montes', is referred to in several testimonies (Peramato, Fragoso, Torreblanca Villalpando, Pérez de Moya, Gómez de Huerta, Fuentelapeña and Lugo y Dávila). The other case, referred to as Magdalena Muñoz, came from the Coronada convent in Úbeda, and in addition to being mentioned in the letters of Fray Agustín de Torres, by Arnauld de Ronsil and in the *Relación de suceso* of 1617, is also taken up by Fuentelapeña and Lugo y Dávila, who differentiate her from María Muñoz. Although Morel d'Arleux, 'Las "Relaciones de Hermafroditas"', p. 268, and Zamora Calvo, '*In virum mutata est*', pp. 438–9, identify them as the same case, Alcalá Galán, 'El andrógino', p. 108, recalls that Lugo y Dávila in his novel *El Andrógino* (1622) differentiates between the two cases. We believe these are two different cases. The discussions by Peramato, Gómez de Huerta and Torreblanca date from 1576, 1599 and 1613, respectively, before the Úbeda case took place. Morel D'Arleux suggests that the date of 1617 must be false. But the characteristics of each case are in fact different. Peramato and Fragoso point out that once María had accepted her change in sex, she was made a priest, a step that did not take place in the case of the girl from Úbeda. Under these circumstances, the girl's father was content to have found an heir.

3. On these women in the Middle Ages, see U. R. Hotchkiss, *Clothes Make the Man: Female Cross Dressing in Medieval Europe* (New York: Garland, 1996). For the modern period, there is an abundance of materials, particularly for Holland and France. See R. M. Dekker and L. C. Van de Pol, *The Tradition of Female Transvestism in Early Modern Europe* (London: Macmillan, 1989), and Steinberg, *La Confusion des Sexes*. Dekker and Van de Pol have worked on a corpus of 119 cases taken from the Dutch penal archives, although they do try to cover other European countries. They mention the case of the Spaniard Catalina de Erauso (Dekker and Van de Pol, *The Tradition of Female Transvestism*, p. 114), as does Steinberg, *La Confusion des Sexes*, pp. 77–8. We should also mention the cross-dressed Juliana de los Cobos (who took part in the wars in Italy by fighting in the troops of Carlos V), and the cases of a Basque woman and an Asturian woman who

dressed as men in the sixteenth century in order to save their husbands; see F. Delpech, '"Muger hay en la guerra": Remarques sur l'exemplaire et curieuse carrière d'une guerrière travestie, Juliana de los Cobos', in A. Redondo (ed.), *Relations entre hommes et femmes en Espagne aux XVIe et XVIIe siècles* (Paris: Presses de l'Université de la Sorbonne, 1995), pp. 53–65; Velasco, *The Lieutenant Nun*, pp. 33–4; and Aresti Esteban, 'The Gendered Identities', p. 404. Similar cases are those of the female bandits Doña Victoria de Acevedo and Doña Josefa Ramírez y Espinela. Furthermore, the cases of some nuns, following the example of the saints in the Golden Legend who decided to dress as men, are worth a mention. Juana Inés de la Cruz at the beginning of the seventeenth century is a case in point. She asked her mother for permission to dress as a man and study at the University of Mexico. Also, there is the case of María de San Antonio, a nun who remained for five years in a Franciscan convent dressed as a man before her death (Steinberg, *La Confusion des Sexes*, p. 69). As these cases indicate a temporary or provisional transformation, they are not elaborated upon here.

4. Gilbert, *Early Modern Hermaphrodites*, pp. 2, 5, 9.
5. Zamora Calvo, '*In virum mutata est*', p. 439, alludes to this correlation as a possible cause of the reform of canon law with the objective of excluding hermaphrodites from the religious orders, except in some unusual cases that will be referred to later on. However, the formulation of such criteria in texts by Martín de Azpilcueta, for example, precedes the majority of cases that are mentioned.
6. On the military question and the nineteenth century, see R. Cleminson and F. Vázquez García, 'The Hermaphrodite, Fecundity and Military Efficiency: Dangerous Subjects in the Emerging Liberal Order of Nineteenth-Century Spain', in K. Fisher and S. Toulalan (eds), *Bodies, Sex and Desire from the Renaissance to the Present* (Basingstoke: Palgrave Macmillan, 2011), pp. 70–86.
7. Dekker and Van de Pol point out that of 93 women (out of 119 in Holland, mainly from the seventeenth and eighteenth centuries) whose male professions are known, 83 of them were in the army or navy at a given moment (Dekker and Van de Pol, *The Tradition of Female Transvestism in Early Modern Europe*, pp. 9–10). On similar circumstances in France, where there was a tradition of 'warrior saints', see Steinberg, *La Confusion des Sexes*, pp. 55–90. On the literary references to the military woman and the case of 'María la Bailaora' in Lepanto, see Morel d'Arleux, 'Las "Relaciones de Hermafroditas"', p. 267. On the tradition of the *serranas* and other warrior women, brought back to life in the Golden Age by Vélez de Guevara y Lope, see Vázquez García and Moreno Mengíbar, *Sexo y Razón*, pp. 390–400, and del Río Parra, *Una era de monstruos*, p. 100. On women dressed as soldiers in the Spanish *comedias* of Lope and others, see Bravo-Villasante, *La Mujer Vestida de Hombre*, pp. 40–5.
8. We thank Chema Fraile of the University of Cadiz for bringing this case to our attention.
9. On the practice of the *enquête* and its relationship to the creation of the modern state, see M. Foucault, *La verdad y las formas jurídicas* (Barcelona: Gedisa, 1980), pp. 82–5.
10. F. J. Campos y Fernández de Sevilla, *Las Relaciones Topográficas de Felipe II. Índices, Fuentes y Bibliografía* (El Escorial: Real Centro Universitario Escorial María Cristina, 2007), p. 468, at http://www.rcumariacristina.com/ficheros/JavierCampos [accessed 4 May 2013].
11. C. Viñas y Mey and R. Paz, *Relaciones Histórico-Geográfico-Estadísitcas de los pueblos de España hechas por iniciativa de Felipe II. Provincia de Madrid* (Madrid: CSIC/Instituto Balmes de Sociología, 1949), pp. 630–1.
12. Ibid., p. 231.
13. Ortiz, 'From Hegemony to Subordination', and Velasco, *Male Delivery*, pp. 56–79.

14. Marchetti, *L'invenzione della bisessualità*, pp. 119–20.
15. For example Park, 'The Rediscovery of the Clitoris', p. 183.
16. Cf. Marchetti, *L'invenzione della bisessualità*, pp. 242–3.
17. Bourdieu, *Sur l'État*, p. 290.
18. A classic analysis of this plurality and mix in contrast to the centralizing thesis is A. M. Hespanha, *Vísperas del Leviatán. Instituciones y poder político (Portugal, siglo XVII)* (Madrid: Taurus, 1989), pp. 363–89.
19. Elisabeth Perry and Federico Garza Carvajal have remarked upon this warrior ethos when explaining the acceptance of sex transformation in the case of Catalina de Erauso: Perry, 'From Convent to Battlefield', pp. 412–13, and Garza Carvajal, *Quemando Mariposas*, pp. 34–40. In Spain, the ideal of civic humanism ('civilidad'), encapsulated by Castiglione's *El Cortesano*, was not established in the first half of the sixteenth century, given the influence of the feudal model, which was in turn based on the exercise of arms; see F. Ampudia de Haro, *Las Bridas de la Conducta. Una aproximación al proceso civilizatorio español* (Madrid: CIS, 2007), pp. 38–9. However, in the first third of the seventeenth century, this model was already considered to be passé in light of the prudence exercised by the new courtly identities (ibid., pp. 53–7).
20. On these changes with respect to warriors, see N. Elias, *The Civilizing Process: Sociogenetic and Psychogenetic Investigations*, trans. E. Jephcott (Malden, Oxford and Carlton: Blackwell, 2000), pp. 387–97. On the characteristics of courtly society, see Elias, *The Court Society*, pp. 66–77.
21. Cartagena-Calderón, '"Él es tan rara persona"', p. 147.
22. On developments in European military technology and tactics, see G. Parker, *The Military Revolution: Military Innovation and the Rise of the West, 1500–1800* (Cambridge: Cambridge University Press, 1988), and A. Campillo, *La fuerza de la razón. Guerra, Estado y ciencia en el Renacimiento* (Murcia: Edit.um, 2008), pp. 23–35.
23. On the ideal of humanistic civility as represented by Castiglione's *El Cortesano* in Spain, see n. 19 above.
24. Rappaport, 'Mischievous Lovers', pp. 13–14.
25. I. Burshatin, 'Elena alias Eleno. Genders, Sexualities and "Race" in the Mirror of Natural History in Sixteenth-Century Spain', in S. P. Ramet (ed.), *Gender Reversals and Gender Cultures: Anthropological and Historical Perspectives* (New York: Routledge, 1996), pp. 105–22, on pp. 115–17.
26. Garza Carvajal, *Quemando Mariposas*, pp. 189–253; O. González Gómez, 'Los discursos sobre la sodomía en las Historias de la conquista de América' (1007), at http://posgradoestudioslatinoamericanos.blogspot.com.es/2009_12_01_archive.html [accessed 11 April 2013]; F. Molina, 'La sodomía a bordo. Sexualidad y poder en la Carrera de Indias', *Revista de Estudios Marítimos y Sociales*, 3:3 (2010), pp. 9–20; F. Molina, 'Crónicas de la sodomía. Representaciones de la sexualidad indígena a través de la literatura colonial', *Bibliographica Americana. Revista Interdisciplinaria de Estudios Coloniales*, 6 (2010), pp. 1–12.
27. Garza Carvajal, *Quemando Mariposas*, pp. 49–54, remarks that 'effeminacy' and 'passivity' were attributes that were given to all the inhabitants in the Indies from the beginning of the sixteenth century. But he also argues that this early variety of 'effeminate sodomy' in the Indies was produced in the middle of the seventeenth century in response to the 'decadent domination' of the Spaniards.
28. Rappaport, 'Mischievous Lovers', p. 9.
29. We must avoid any notion of essentialism with respect to identities, if we can speak of such, within the *Ancien Régime*. The notion of 'oneself', to draw on the theoretical appara-

tus provided by P. Ricoeur, *Oneself as Another*, trans. K. Blamey (Chicago, IL and London: University of Chicago Press, 1992), pp. 2–3, is less of an *idem* or 'selfness' than an *ipse*, configured by the actions that respond to interpellation by others. In this way, caution must be exercised when assuming 'pre-modern' identities, which would be supposedly static, fixed and constructed, in opposition to a modern identity, defined as a 'project'.

30. On this concept, see S. Greenblatt, *Renaissance Self-Fashioning: From More to Shakespeare* (Chicago, IL: University of Chicago Press, 1980).

31. If the pre-modern subject cannot be understood in essentialist terms, neither can it be viewed from a constructivist position, that is, as a self that constructs its own realities. An excellent critique of the voluntaristic constructivism, cast in the light of recent neo-liberal debates on the 'ownership' of oneself, can be found in the work by E. Burgos, 'Deconstrucción y subversión', in P. Soley-Beltran and L. Sabsay (eds), *Judith Butler en disputa. Lecturas sobre la performatividad* (Madrid and Barcelona: Egales, 2012), pp. 101–34, esp. pp. 127–32, and L. Sabsay, 'De sujetos performativos, psicoanálisis y visiones constructivistas', in P. Soley-Beltran and L. Sabsay (eds), *Judith Butler en disputa. Lecturas sobre la performatividad* (Madrid and Barcelona: Egales, 2012), pp. 135–68, especially pp. 142–50.

32. G. Folch Jou and M. S. Muñoz Calvo, 'Un pretendido caso de hermafroditismo en el siglo XVI', *Boletín de la Sociedad de Historia de la Farmacia*, 93 (1973), pp. 20–33; E. Maganto Pavón, 'La intervención del Dr. Francisco Díaz en el proceso inquisitorial contra Elena/o de Céspedes, una cirujana transexual condenada por la Inquisición de Toledo en 1587', *Historia de la Urología Española*, 60:8 (2007), pp. 873–86; Maganto Pavón, *El proceso inquisitorial*.

33. Escamilla, 'A propos d'un dossier inquisitorial'. The work by Zamora Calvo, '*In virum mutata est*', pp. 440–2, alleging 'transsexuality', is not devoid of this medicalizing approach. Maganto Pavón, *El proceso inquisitorial*, pp. 9–10, argues in contrast to Escamilla that this was not a case of intersexuality but transexualism. The volume is an attempt to explore 'los aspectos estrictamente médicos o psicopatológicos de la protagonista' (Maganto Pavón, *El proceso inquisitorial*, pp. 7–8). The problem with these studies is that their attempt to interpret past experiences in light of today's categories obfuscates the peculiarities of the period referred to.

34. J. Pérez Escohotado, *Sexo e Inquisición en España* (Madrid: Temas de Hoy, 1992), pp. 158–60; J. Eslava Galán, *Historia secreta del sexo en España* (Barcelona: Planeta, 1992), p. 78; and A. Sánchez Vidal, *Esclava de Nadie* (Madrid: Espasa Calpe, 2010).

35. Barbazza, 'Un caso de Subversión Social', esp. pp. 34, 39.

36. Burshatin, 'Written on the Body', p. 429, proclaims the lesbianism of Céspedes and critiques those studies that are overly obsessed with the 'incommensurability' of different historical periods and unwilling to utilize similar categories in order to understand the past. S. Velasco, *Lesbians in Early Modern Spain* (Nashville, TN: Vanderbilt University Press, 2011), pp. 68–9, also refers to the lesbianism of Céspedes, but with a greater degree of historical caution.

37. Burshatin, 'Written on the Body', p. 422.

38. Burshatin, 'Elena alias Eleno', p. 106, and Velasco, *Male Delivery*, p. 104.

39. Vázquez García and Moreno Mengíbar, 'Un solo sexo', esp. pp. 99–103; Vázquez García and Moreno Mengíbar, *Sexo y Razón*, pp. 191–6; and Kagan and Dyer, *Inquisitorial Inquiries*, pp. 53–9.

40. Kagan and Dyer, *Inquisitorial Inquiries*, p. 56, argue that the continual identity transitions in Céspedes reflect the mobility of artesanal and commercial relations in early modernity in contrast to the preconceived notion of pre-modern life as primarily sedentary.

41. F. Vázquez García, *Tras la autoestima. Variaciones sobre el yo expresivo en la modernidad tardía* (San Sebastián: Gakoa, 2005), p. 97.

42. G. H. Mead, 'La genesis del *self* y el control social', *Revista Española de Investigaciones Sociológicas*, 55 (1991), pp. 165–86. The article was originally published in 1925 in the *International Journal of Ethics*.

43. On the 'ambiguous' status of the accusation of bigamy, given the fact that the officers of the Inquisition were aware that Céspedes's first husband had died before she married María del Caño, see Maganto Pavón, 'La intervención', p. 189.

44. Butler, *Bodies that Matter*, p. 12.

45. Kagan and Dyer, *Inquisitorial Inquiries*, p. 39 n. 6.

46. Butler, *Gender Trouble*, p. 140

47. Burshatin, 'Written on the Body', p. 434.

48. On the connections between the life of Céspedes and the picaresque, see ibid., p. 422. On the construction of the 'picaresque self' by means of the autobiographical account during the crisis of organicist feudalism, see J. C. Rodríguez, *La literatura del pobre* (Granada: Comares, 1994), pp. 116–45. On social mobility in the *Ancien Régime*, see Kagan and Dyer, *Inquisitorial Inquiries*, p. 56, and Ampudia de Haro, *Las Bridas de la Conducta*, p. 3.

49. Burshatin, 'Written on the Body', pp. 423–4, notes that sodomy was under the jurisdiction of the civil authorities in Castile, not the Inquisition.

50. Ibid., pp. 432–3. On Francisco Díaz, see López Piñero et al., 'Díaz, Francisco', in *Diccionario Histórico de la Ciencia Moderna en España*, pp. 278–81. Díaz was well informed on the nature of the genital organs. In 1588 he published his *Tratado nuevamente impresso, de todas las enfermedades de los Riñones, Vexiga y Carnosidades de la Verga y Urina* (Madrid: Francisco Sánchez).

51. 'y se le hicieron allí unas grietas por donde muchos días andubo destilando sangre y se le enmustió el dicho miembro bolviéndosele como de esponja y ésta le fue cortando poco a poco de manera que a benido a quedar sin ello' (and they made some cuts there from which she bled over many days and the said member dried up and became a kind of sponge and this was cut back little by little so that she came not to have it all); *Proceso Inquisitorial de Elena o Eleno de Céspedes*, Archivo Histórico Nacional, Inquisición, Legajo 234, n° 24.

52. During the trial it emerged that Elena possessed a library with extensive works on natural history and medicine in Spanish and Latin. It is possible that Elena cited Pliny before the court to defend the unusual nature of her case (Kagan and Dyer, *Inquisitorial Inquiries*, pp. 46–7 n. 29). On the impressive culture shown by Elena, see Burshatin, 'Written on the Body', p. 438.

53. It is possible that the Valencian surgeon who trained Elena during one of her stays in Madrid belonged to this group of *morisco* physicians called 'sanadores'. On *morisco* medicine, see García-Ballester, *Los Moriscos y la Medicina*, pp. 60–136.

54. For a perceptive comparison between the two, see S. Velasco, 'Interracial Lesbian Erotics in Early Modern Spain: Catalina de Erauso and Elena/o de Céspedes', in L. Torres and I. Perpetusa-Seva (eds), *Tortilleras: Hispanic and US Latina Lesbian Expression* (Philadelphia, PA: Temple University Press, 2003), pp. 213–27, and Velasco, *Lesbians in Early Modern Spain*, pp. 78–81.

55. On this point, see Barbazza, 'Un caso de Subversión Social', p. 19. On the vagabond, see Castel, *Les métamorphoses de la question sociale*, pp. 40–3.

56. Barbazza, 'Un caso de Subversión Social', p. 20; Burshatin, 'Written on the Body', pp. 436–8.

57. F. Vázquez García, *La invención del racismo. Nacimiento de la biopolítica en España, 1600–1940* (Madrid: Akal, 2009), pp. 110–20.

58. Burshatin, 'Written on the Body', pp. 425, 442–3.

59. Burshatin, 'Elena alias Eleno', pp. 115–17, focuses on this point.

60. Burshatin, 'Written on the Body', p. 429.

61. R. Carrasco and B. Vicent, 'Amours et marriage chez les morisques au XVᵉ siècle', in A. Redondo (ed.), *Amours Légitimes et Illégitimes en Espagne (XVIe–XVIIe siècles)* (Paris: Publications de la Sorbonne, 1985), pp. 133–50, esp. pp. 144–5; E. Rudelle-Berteaud, 'Divergencias moriscas y cristianas sobre erotismo y afectividad', *Al-Andalus*, 248 (2004), at http://old.webislam.com/numeros/2004/248/temas/divergencias_moriscas_cristianas.htm [accessed 4 May 2013]; Velasco, *Male Delivery*, pp. 117–19; and T. Mantecón Movellán, 'Los mocitos de Galindo: sexualidad *contra natura*, culturas proscritas y control social en la edad moderna', in T. Mantecón Movellán (ed.), *Bajtín y la historia de la cultura popular* (Santander: Universidad de Cantabria, 2008), pp. 209–40, esp. pp. 218–19. However, G. S. Hutcheson, 'The Sodomitic Moor: Queerness in the Narrative of *Reconquista*', in G. Burger and S. F. Kruger (eds), *Queering the Middle Ages* (Minneapolis, MN and London: University of Minnesota Press, 2001), pp. 99–122, downplays the evidence for this, locating the association more concretely in the nineteenth century.

62. Marchetti, *L'invenzione della bisessualità*, pp. 242–4.

63. On the use of luxurious dildos ('caralhos') in the Galaico-Portuguese *cantigas* of the lower Middle Ages by libidinous nuns, see E. Lacarra Lanz, 'Homoerotismo femenino en los discursos normativos medievales', in A. Chas Aguión and C. Tato García (eds), *Siempre soy quien ser solía: Estudios de literatura española medieval en homenaje a Carmen Parrilla* (A Coruña: Universidade da Coruña, 2009), pp. 205–28. This author has researched the use of *gedomichés* in a range of Spanish medieval sources from the *Speculum al foderi* to Eximenis. On the use of the dildo in the Golden Age, see Velasco, *Lesbians in Early Modern Spain*, pp. 35–41, 55–6 and 169–71.

64. Maganto Pavón, *El proceso inquisitorial*, p. 90, and Velasco, *Lesbians in Early Modern Spain*, p. 78.

65. Marchetti, *L'invenzione della bisessualità*, pp. 145–7. On the perception of practical and academic medicine on women with large clitorises, see J. Brown, *Sor Benedetta, entre sainte et lesbienne* (Paris: Gallimard, 1986), pp. 16–17, and E. Donoghue, 'Imagined More than Women: Lesbians as Hermaphrodites, 1671–1766', *Women's History Review*, 2:2 (1993), pp. 199–216, esp. pp. 203–10.

66. The majority of commentators discuss laws on female sodomy by drawing on the gloss on Law LXXX, written up in Toro, as contained in the book by the Castilian Antonio Gómez, *Ad leges tauri commentarius* (1616). See Tomás y Valiente, 'El Crimen y Pecado contra Natura', pp. 47–9; Marchetti, *L'invenzione della bisessualità*, pp. 276–8; Velasco, *Lesbians in Early Modern Spain*, pp. 17–30; and F. Garza Carvajal, *Las Cañitas. Un proceso por lesbianismo a principios del XVII* (Palencia: Simancas Ediciones, 2012), p. 54. However, another important legal figure of the sixteenth century, Gregorio López, in his commentary on *Las siete partidas*, argued for the death penalty for tribades. See Brown, *Sor Benedetta*, p. 19.

67. In *El sueño de la viuda* one of the female protagonists undergoes a sexual metamorphosis and a penis is produced. Although at the end of the work it is discovered that this was a dream, the 'sex change' gives rises to a series of amusing amorous adventures throughout the story. See J. I. Díez Fernández, *La poesía erótica de los Siglos de Oro* (Madrid: Ediciones del Laberinto, 2003), pp. 241–3, and Alcalá Galán, 'El andrógino', p. 121. On the hagiographic inspiration of this comedy, see J. Sanz Hermida, 'Ensoñación y transformismo: la parodia erótica en *El sueño de la viuda*, de Fray Melchor de la Serna', in *Studia Aurea. Actas del Congreso de la AISO*, vol. 1 (Toulouse and Pamplona: Centro Virtual Cervantes, 1996), pp. 513–23, on p. 517.

68. Velasco, *Lesbians in Early Modern Spain*, pp. 137–43 and 152–7. Velasco, *Lesbians in Early Modern Spain*, pp. 38–9, and Garza Carvajal, *Las Cañitas*, p. 54, refer to a question elevated by the Aragonese inquisitors to the Supreme Tribunal in 1560 on the subject of whether genital contact and effusion of semen between women, but with no penetration, should be punished. The inquisitors of Madrid judged that such an act did not constitute sodomy and, for this reason, fell outside the Inquisition's jurisdiction.

69. A sentence was given by the royal audiences of Castille and León in 1609, and Inés Santa Cruz was given four hundred lashes and condemned to prison for six years for public scandal. After this, she would be permanently expelled from her locality. This process has been studied and transcribed by Garza Carvajal, *Las Cañitas*. Velasco, *Lesbians in Early Modern Spain*, pp. 39–40, also refers to an inquisitorial trial in Aragón (1656) for sodomy between Ana Aler and Mariana López. These two women were sentenced to two hundred lashes and eight years of exile and were prohibited from living in the same locality. However, in this case, there was no evidence given to suggest that the women had used any instrument for the purposes of penetration. The penal expert Antonio Gómez refers to a case of two Spanish nuns who were found guilty of penetration with an instrument, but gives no more details apart from the fact that they were burned alive (Tomás y Valiente, 'El Crimen y Pecado contra Natura', p. 48).

70. See Jacquart and Thomasset, *Sexualidad y Saber Médico*, p. 149, on the figure of the hermaphrodite as a privileged source of knowledge on the female body. See also the reflection on 'androginia y taumaturgia' in Libis, *El Mito del Andrógino*, pp. 108–13.

71. Kagan and Dyer remark: 'a great many Spaniards ... perceived her as a kind of miracle worker and flocked to the hospital in droves to be cured by her' (Kagan and Dyer, *Inquisitorial Inquiries*, p. 57).

72. J. Gómez de Huerta, *Historia Natural de Cayo Plinio Segundo* (Madrid: Iván González, 1629), p. 262, and Fuentelapeña, *El Ente Dilucidado*, pp. 244–5.

73. The reliability of the copy (made by Juan Bautista Muñoz in 1784 from another copy drawn up in the eighteenth century) of the manuscript from which details of the autobiography of Catalina de Erauso are inferred has been placed in doubt by Menéndez Pelayo. For this reason, instead of using this version, whose most recent editions are R. de Vallbona (ed.), *Vida i sucesos de la Monja Alférez: Autobiografía atribuida a Doña Catalina de Erauso* (Tempe: Center for Latin American Studies, 1992), and the English translation by M. Stepto and G. Stepto (eds), *Lieutenant Nun: A Memoir of a Basque Transvestite in the New World: Catalina de Erauso* (Boston, MA: Beacon Press, 1996), we favour that of Pedro Rubio Merino. This text is based on the two manuscripts by Catalina de Erauso herself. They were discovered by Rubio in the Archivo de la Catedral de Sevilla. This edition also includes the transcription of the whole documentary dossier on Catalina de Erauso conserved in the Archivo General de Indias. See P. Rubio Merino (ed.), *La Monja Alférez Doña Catalina de Erauso. Dos Manuscritos Autobiográficos Inéditos* (Seville: Edi-

ciones del Cabildo Metropolitano de la Catedral de Sevilla, 1995). The most extensive biography, which includes an excellent survey of Catalina's family background, is still J. I. Tellechea Idígoras, *La Monja Alférez, Doña Catalina de Erauso* (San Sebastián: Sociedad Guipuzcoana de Ediciones y Publicaciones, Obra Cultural de Kutxa, 1992). See also C. R. Boxer, *Mary and Misogyny: Women in Iberian Expansion Overseas 1415–1815* (London: Duckworth, 1975); Perry, 'From Convent to Battlefield'; Velasco, *The Lieutenant Nun*; Garza Carvajal, *Quemando Mariposas*, pp. 32–64.

74. Perry, 'From Convent to Battlefield', p. 411, subscribes to this interpretation and erroneously, in our view, suggests a dichotomous and essentialist frame for the sex/gender regime of the Golden Age. Her analysis does not take into account the decisive contribution by Laqueur. Although Velasco, *The Lieutenant Nun*, p. 9, does not either, this is rectified in Velasco, *Male Delivery*, p. 104, where Laqueur's insights are present. It is Garza Carvajal, *Quemando Mariposas*, pp. 34–6, 54–9, 259–60, who has insisted most, by means of a post-colonial analysis (drawing on Said, Spivak, Sangari, Sarkar and Sinha), that Erauso, who took on the name Alonso Díaz, was an incarnation of manliness concordant with imperial Spain.

75. See Aresti Esteban, 'The Gendered Identities', and F. Vázquez García and R. Cleminson, 'Subjectivities in Transition: Gender and Sexual Identities in Cases of "Sex Change" and "Hermaphroditism" in Spain, *c.* 1500–1800', *History of Science*, 48:159 (2010), pp. 1–38. The reference to the 'inner moustache' of Erauso comes from Aresti Esteban, 'The Gendered Identities', p. 406.

76. Velasco, *The Lieutenant Nun*, p. 67.

77. On Catalina de Erauso as an exemplar of the 'perfect' Spanish man and woman, see Garza Carvajal, *Quemando Mariposas*, p. 38.

78. Miguel de Erauso, the father of Catalina, and one of her brothers were captains. The captain of the ship that took her to America the first time was Catalina's uncle, Esteban Ciguino. The prioress of the Dominican convent of San Sebastián was a cousin of her mother's, Dª Úrsula de Unzá y Sarasti.

79. An examination of the documentation in the Archivo General de Indias shows the variety of military personnel that Catalina de Erauso was able to get to testify in her favour for her pension. The mobilization of this social capital can only be explained by the links that her family had with the army. A discussion of these personalities can be found in 'Memorial de la Monja Alférez, Doña Catalina de Erauso con la relación de sus méritos y servicios', in Rubio Merino, *La Monja Alférez*, p. 136. Other sources of support came from her use of the Basque language and her close ties with *corregidores* and authorities at home in order to secure her release from prison or her immunity from the law (ibid., pp. 40–1).

80. The insertion of Catalina into the military and colonial *habitus* of the period, with special reference to her 'autobiography' and to what the author understands as the playing down of her transgressive role in an explicit sense but with certain hints to the seventeenth-century reader, is discussed in A. Borrachero Mendíbil, 'Catalina de Erauso ante el patriarcado colonial: un estudio de *Vida i sucesos de la Monja Alférez*', *Bulletin of Hispanic Studies*, 83:6 (2006), pp. 485–95.

81. On the 'courtization' of warriors and the contrast between the court nobility and the warrior nobility, see Elias, *The Civilizing Process*, pp. 387–97. This model has been applied to the Spanish case in the excellent study by Ampudia de Haro, *Las Bridas de la Conducta*, pp. 56–66. The conversion of the Spanish court into the European model began to take place in 1561 with the establishment of the court of Felipe II in Madrid (ibid., p. 54).

82. Perry, 'From Convent to Battlefield', p. 412. Both Elisabeth Perry and Federico Garza (Garza Carvajal, *Quemando Mariposas*, pp. 34–8) emphasize the identification of Catalina with the warrior ethos of the Reconquest and the American conquest. Both authors, however, emphasize the gender and postcolonial analysis and fail to evaluate the social capital that Catalina possessed in the process of her construction as a man. In order to understand these family ties, the work by Tellechea Idígoras, *La Monja Alférez*, is essential.

83. Rubio Merino, *La Monja Alférez*, p. 11.

84. One of the best works on the culture of the duel and the formation of male identity is R. Nye, *Masculinity and Male Codes of Honor in Modern France* (New York and Oxford: Oxford University Press, 1993).

85. H. Clementi, *La Frontera en América. Una clave interpretativa de la historia americana* (Buenos Aires: Leviatán, 1985); A. Grimson (ed.), *Fronteras, Naciones e Identidades. La periferia como centro* (Buenos Aires: CICCUS-La Crujía, 2000); F. Opere, *Historias de la frontera: el cautiverio en la América hispánica* (Buenos Aires: FCE, 2001).

86. On the relationship between civilizing control and bellicosity in the construction of the State, see Elias, *The Civilizing Process*, pp. 191–4.

87. Cartagena-Calderón, '"Él es tan rara persona"', p. 144.

88. A. Gutiérrez Nieto, 'El pensamiento económico político y social de los arbitristas', in M. Menéndez Pidal (ed.), *Historia de España. El Siglo del Quijote*, vol. 1 (Madrid: Espasa Calpe, 1986), pp. 233–351, on pp. 233–5.

89. J. A. Maravall, *La Cultura del Barroco* (Barcelona: Ariel, 1983), pp. 93–4; Cartagena-Calderón, '"Él es tan rara persona"', pp. 139–40; Velasco, *Male Delivery*, pp. 108–9; A. L. Martín, 'Sodomitas, putos, doncellos y maricotes en algunos textos de Quevedo', *La Perinola*, 12 (2008), pp. 107–22; Vázquez García, *La invención del racismo*, pp. 123–6.

90. Maravall, *La Cultura del Barroco*, pp. 94–6; Velasco, *Male Delivery*, pp. 197–9.

91. R. González Cañal, 'El lujo y la ociosidad durante la privanza de Olivares: Bartolomé Jiménez Patón y la polémica sobre el guardainfante y las guedejas', *Criticón*, 53 (1991), pp. 71–96, esp. pp. 73–4, and Velasco, *Male Delivery*, pp. 107–8.

92. 'hoy no parece bien un abuso tan indigno [las guedejas y copetes] del brío español, que pervierte las acciones naturales y hace a sus dueños hermafroditas'; Gómez Tejada, *León Prodigioso* (1636), cited in González Cañal, 'El lujo', p. 87.

93. Rappaport, 'Mischievous Lovers', p. 15. In a similar vein, see Alcalá Galán, 'El andrógino', p. 111.

94. Velasco, *Male Delivery*, p. 43. In the transformation from man to woman, except in the case of a prenatal change as signalled by Huarte de San Juan, human artifice intervened. In the reverse case, change was part of the natural order that tended towards perfection. Bovistuau, whose treatise on marvels was translated in 1603, discussed Nero and Heliogabalus as examples of such voluntary alterations. See del Río Parra, *Una era de monstruos*, p. 98, and Velasco, *Male Delivery*, pp. 96–9. These transformations possessed a counter-natural significance and as such were represented in the theatre as being caused by spells and witchcraft, as in *La hechicera* (*c.* 1637) by Quiñones de Benavente and in *Los Putos* (*c.* 1645) by Jerónimo de Cáncer. On this subject, see Restrepo-Gautier, 'Afeminados', p. 210, and Martínez, 'Mari(c)ones, travestis y embrujados', p. 27. Velasco, *Male Delivery*, p. xiv, also refers ro a 'relación de suceso' from 1606 recording a man who became pregnant as the result of a witch's spell.

95. Although there are exceptions, for example in her discussion of the well-known work by María Zayas, *Mal presagio: casar lejos*, Velasco, *Male Delivery*, p. 115, in other sec-

tions the author engages in a somewhat mechanical association between effeminacy and sodomy, for example in ibid., p. 97. A similar approach is taken by Martín, 'Sodomitas'.

96. In contrast to Velasco, *Male Delivery*, and the tendency to assimilate sodomy and effeminacy, see Mantecón Movellán, 'Los mocitos de Galindo', p. 221, and Martínez, 'Mari(c) ones, travestis y embrujados', p. 27.

97. Garza Carvajal, *Quemando Mariposas*, pp. 189–253. M. E. Perry, *Ni espada rota ni mujer que trota. Mujer y desorden social en la Sevilla del Siglo de Oro* (Barcelona: Crítica, 1993), pp. 124–7, also refers to groups of effeminates who practised sodomy in Seville in the closing years of the sixteenth century, although she points out that not all those who adopted such an appearance were sodomites.

98. This connection is illustrated by González Cañal, 'El lujo'.

99. Cartagena-Calderón, '"Él es tan rara persona"', pp. 148–9.

100. Vázquez García, *La invención del racismo*, pp. 123–6.

101. On the reinterpretation of the myth of Salmacis and Hermaphrodito in this sense, see Velasco, *Male Delivery*, p. 32.

102. On the favourable reception of stoicism and Christian thought drawing on Seneca in the circle around the Conde Duque de Olivares, which helped to fuse the old austerity and self-restraint of the aristocracy, see Vázquez García, *La invención del racismo*, p. 123. On neo-stoicism in Spanish political thought in the 1600s, see G. Oestreich, *Neostoicism and the Early Modern State* (Cambridge: Cambridge University Press, 1982), pp. 102–4; J. A. Fernández-Santamaría, *Razón de Estado y Política en el Pensamiento Español del Barroco (1595–1640)* (Madrid: Centro de Estudios Constitucionales, 1986), pp. 83–4; and J. A. Maravall, *Teoría del Estado en España en el siglo XVII* (Madrid: Centro de Estudios Constitucionales, 1997), pp. 106–7.

103. P. Fernández Navarrete, *Conservación de Monarquías y Discursos Políticos* (1626; Madrid: Instituto de Estudios Fiscales, Ministerio de Hacienda, 1982), p. 92.

104. On the critique of court life in the times of the Duque de Lerma and the specific connotations with respect to (not just economic) corruption, see A. Feros, *El Duque de Lerma. Realeza y privanza en la España de Felipe III* (Madrid: Marcial Pons, 2009), pp. 323–35.

105. Vázquez García, *La invención del racismo*, p. 122. On the programme of the Junta de Reformación, see Maravall, *La Cultura del Barroco*, p. 95; M. Martín Rodríguez, *Pensamiento Económico Español sobre la Población* (Madrid: Ediciones Pirámide, 1984), pp. 261–5; González Cañal, 'El lujo'; M. Jiménez Monteserín, *Sexo y bien común. Notas para la historia de la prostitución en España* (Cuenca: Ayuntamiento de Cuenca, Instituto 'Juan de Valdés', 1994), pp. 158–63; and Cartagena-Calderón, '"Él es tan rara persona"', p. 152.

106. González Cañal, 'El lujo', p. 87.

107. Velasco, *The Lieutenant Nun*, pp. 60–70.

108. González Cañal, 'El lujo'; Cartagena-Calderón, '"Él es tan rara persona"'; Restrepo-Gautier, 'Afeminados'; Velasco, *Male Delivery*; Martín, 'Sodomitas'; Alcalá Galán, 'El andrógino'; and Martínez, 'Mari(c)ones, travestis y embrujados'.

109. Cartagena-Calderón, '"Él es tan rara persona"', pp. 162–7.

110. The work by Quevedo, as Restrepo-Gautier, 'Afeminados', pp. 206–7, has argued, is particularly subversive because his *marión* is not completely effeminized; his demeanour is womanly but his appearance is not – no make-up or hair curls are employed. Rather than questioning the border between the sexes, however, this character shores up difference within the hierarchy of rank.

111. Ibid., pp. 205–8; Velasco, *Male Delivery*, p. 112; and Martínez, 'Mari(c)ones, travestis y embrujados', pp. 22–4.

112. Restrepo-Gautier, 'Afeminados', pp. 203–4. Alcalá Galán, 'El andrógino', p. 116, refers to *La Gran Sultana*. This work by Cervantes includes as part of its intrigue a false 'improvement by sex' by a supposed 'hidden hermaphrodite'.
113. Velasco, *Male Delivery*, pp. 28–49, and Martínez, 'Mari(c)ones, travestis y embrujados', pp. 24–6.
114. Martínez, 'Mari(c)ones, travestis y embrujados', p. 19, and Alcalá Galán, 'El andrógino', pp. 116–17.
115. For different interpretations on this point, see Restrepo-Gautier, 'Afeminados', pp. 207–8, and Velasco, *Male Delivery*, pp. 124–6.
116. Here we follow the argument of J. I. Díez Fernández, *Tres discursos de mujeres. (Poética y hermenéutica cervantinas)* (Alcalá de Henares: Centro de Estudios Cervantinos, 2004), pp. 152–3.
117. Martínez, 'Mari(c)ones, travestis y embrujados', p. 14.
118. The Cortes de Valladolid approved in 1523 an order prohibiting the use of masks and disguises and any other device that would prevent someone from being identified. The fine was doubled if this occurred at night. See http://www.mcu.es/archivos/lhe/action.find.jsp# [accessed 30 June 2012].
119. A. de León Pinelo, 'Velos antiguos y modernos en los rostros de las mujeres. Sus conveniencias y daños' (1639), *Lemir*, 13 (2009), pp. 235–388, esp. pp. 373–4.
120. L. Bueno, 'Identidad sexual del personaje en el teatro aurisecular', in A. Ceballos, R. Espejo and B. Muñoz (eds), *El teatro del género. El género del teatro. Las artes escénicas y la representación de la identidad sexual* (Madrid: Fundamentos, 2009), pp. 41–67, argues that a woman dressed as a man was sanctioned by moralists as a result of the eroticism that resulted from the restrictions of the female form in male attire. By contrast, men dressed as women produced a comic efect. It is possible that this interpretation is valid for those comedies that introduce this motif, such as *El ruiseñor de Sevilla* (Lope de Vega), *Mudarse por mejorarse* (Alarcón) or *La república al revés* (Tirso de Molina). But in the case of short theatre plays, as analysed by Restrepo-Gautier, 'Afeminados', Velasco, *Male Delivery*, Martín, 'Sodomitas', and Martínez, 'Mari(c)ones, travestis y embrujados', such an act, presented as comic, constituted a critique of the decay of masculinity in the empire and, at times, the preference for sodomitic behaviour.
121. See Alcalá Galán, 'El andrógino', pp. 110–11. Velasco, *Male Delivery*, p. 160, refers to a recommendation by the Consejo de Castilla in 1600, which pointed out that although it was not fitting that women acted in the theatre, it was less harmful than men dressed as women. On this question, see González Cañal, 'El lujo', pp. 93–4, and Velasco, *The Lieutenant Nun*, pp. 35–42.
122. As Díez Fernández, *Tres discursos de mujeres*, pp. 153–4, has emphasized, different ontological orders cannot be confused. Transvestism in daily life does not contain the same significance as on the stage.
123. We follow here the analysis set out by Alcalá Galán, 'El andrógino'. See also Martínez, 'Mari(c)ones, travestis y embrujados', pp. 14–15.
124. See Cartagena-Calderón, '"Él es tan rara persona"', p. 140, and Velasco, *Male Delivery*, pp. 110–11.
125. See Velasco, *Male Delivery*, p. 115, and Alcalá Galán, 'El andrógino', p. 122.
126. M. Rich Greer, 'María de Zayas and the Female Eunuch', *Journal of Spanish Cultural Studies*, 2:1 (2001), pp. 41–53, on p. 47.
127. P. Bourdieu, *La domination masculine* (Paris: Raisons d'Agir, 1998), p. 38. On masculinity as a kind of 'nobility' among the Kabyles, see ibid., p. 66.

128. Rich Greer, 'María de Zayas', p. 45, and Alcalá Galán, 'El andrógino', p. 45.
129. Velasco, *The Lieutenant Nun*, p. 35.
130. Rubio Merino, *La Monja Alférez*, p. 86. In the nineteenth century, studies such as that of Menéndez Pelayo wrote of Catalina de Erauso's story as legendary, and doctors such as Nicolás León considered her to be 'un seudo hermafrodita hipospádico' (ibid., p. 28). Here we can see the attempt to translate the language of 'rank' of the sexual *Ancien Régime* into the biological terminology of nineteenth-century medicine.
131. The work of M. L. de Padilla Manrique y Acuña, *Excelencias de la Castidad* (Madrid: Biblioteca de Autores Españoles, 1975), was first published in 1642. Here the benefits and virtues associated with virginity are extensively discussed.
132. Perry, 'From Convent to Battlefield', p. 397. Catalina managed to reduce the size of her breasts to almost nothing by means of a kind of corset given to her by an Italian. The wearing of this device was extremely painful (see Rubio Merino, *La Monja Alférez*, p. 17).
133. Perry, 'From Convent to Battlefield', pp. 404–5. An obvious sign of Catalina's popularity was her depiction in the comedy written by Juan Pérez de Montalbán, *La Monja Alférez*, in 1626, published in *Autores dramáticos contemporáneos de Lope de Vega* (Madrid: Biblioteca de Autores Españoles, 1881).
134. Catalina's reluctance to undress when asked to do so by the Alcalde de Potosí so as to submit her to torture relates to her fear of discovery: 'Mandóme que me fuesse desnudando. Hícelo, aunque no de buena gana' (Rubio Merino, *La Monja Alférez*, p. 70).
135. On the religious beliefs of Catalina, see ibid., pp. 43–6.
136. Perry, 'From Convent to Battlefield', p. 413.
137. On the relationship between Catalina and Spanish patriotism, see Rubio Merino, *La Monja Alférez*, pp. 40–3.
138. There is abundant bibliography on these questions. On the offensive against the Spanish clergy's moral relaxation, see Jiménez Monteserín, *Sexo y bien común*, pp. 147–204; H. Kamen, *Cambio cultural en la sociedad del Siglo de Oro. Cataluña y Castilla, siglos XVI–XVII* (Madrid: Siglo XXI, 1998), pp. 259–319; and A. Morgado García, *Ser clérigo en la España del Antiguo Régimen* (Cadiz: Universidad de Cádiz, 2000), pp. 166–72. For the European picture, with numerous references to Iberia, see M. E. Wiesner-Hanks, *Christianity and Sexuality in the Early Modern World: Regulating Desire, Reforming Practice* (London: Routledge, 2000), pp. 101–39.
139. As Morgado García, *Ser clérigo*, p. 168, argues, it took until the early eighteenth century for the effect to be felt.
140. This documentation has been examined extensively by Soyer, 'The Inquisition'.
141. Ibid., p. 551.
142. Ibid., p. 556.
143. For a comparison between these two figures in this respect, see ibid., p. 560.
144. Ibid., p. 562, reminds the reader that the archive of the diocese of Badajoz was destroyed in 1705 during the Peninsular War.
145. A. Rodríguez Sánchez, *Hacerse nadie. Sometimiento, sexo y silencio en la España de finales del siglo XVI* (Lleida: Milenio, 1998).
146. Morel d'Arleux, 'Las "Relaciones de Hermafroditas"', p. 263, and Zamora Calvo, '*In virum mutata est*', p. 439.
147. Marchetti, *L'invenzione della bisessualità*, p. 18.
148. Morgado García, *Ser clérigo*, pp. 52–3.
149. This is not the case in the original Portuguese version, where the physical 'irregularities' are referred to but hermaphroditism is not included. 'Notable deformities' on the face

and hands are mentioned, which could be the lack of eyes or a finger; M. de Azpilcueta, *Manual de confessores et penitentes* (Coimbra: Ioan de Barnyra, 1549), p. 464.

150. M. de Azpilcueta, *Enchiridion sive manuale confessariorum ac poenitentiam* (Paris: Imp. Guillielmi Rovilli, 1587), p. 1081.

151. Marchetti, *L'invenzione della bisessualità*, pp. 19 and 52.

152. Ibid., p. 52.

153. Morel d'Arleux, 'Las "Relaciones de Hermafroditas"', p. 263, and Zamora Calvo, '*In virum mutata est*', p. 439.

154. Marchetti, *L'invenzione della bisessualità*, p. 52.

155. Ibid., p. 19.

156. F. M. de Rodríguez, *Suma de casos de conciencia* (Salamanca: Imp. Diego de Cossío, 1603), p. 512, and Villalobos, *Manual de Confesores*, p. 423.

157. Soyer, 'The Inquisition', p. 559.

3 The Expulsion of the Marvellous: The Decline of the 'One-Sex' Model, 1750–1830

1. Laqueur, *Making Sex*, pp. 19–21. On the reception and critique of Laqueur's ideas among historians of sexuality of eighteenth-century Britain, see K. Harvey, 'The Substance of Sexual Difference: Change and Persistence in Representations of the Body in Eighteenth-Century England', *Gender and History*, 14:2 (2002), pp. 202–23, esp. pp. 203–5.

2. In the case of France in the mid-eighteenth century, with a few exceptions, the thesis that hermaphrodites did not really exist triumphed in medical discourse and in other genres, such as travel literature. On this point, see P. Graille, *Le troisième sexe. Être hermafrodite aux XVIIe et XVIIIe siècles* (Paris: Les éditions arkhê, 2011). Hermaphroditism arose from superstitious accounts of people who possessed 'vices of conformation' in the genitalia. This transitional stage is reflected, for example, in the article by the polygraph C. de Jaucourt, 'Hermaphrodite', in *Encyclopédie de Diderot et D'Alembert*, vol. 15 (1755; Paris: Ed. de Franco Maria Ricci, 1978), fols H27–H28, and continued by Diderot. In the piece by Diderot in the *Supplément à l'Encyclopédie* (1777), only masculine, feminine and asexual varieties were admitted. See J. P. Guicciardi, 'L'hermaphrodite et le prolétaire', *Dix-Huitième Siècle*, 12 (1980), 49–78, esp. p. 64; Foucault, *Les Anormaux*, p. 67; and Graille, *Le troisième sexe*, pp. 58–68. In Britain and colonial America, in part as a result of the influential book by James Parsons, *Mechanical and Critical Inquiry into the Nature of Hermaphrodites* (London: J. Walthoe, 1741), which was in turn drawn upon by de Jaucourt, the idea that hermaphrodites did not exist in the human species also became predominant. See G. S. Rousseau and R. Porter (eds), *Sexual Underworlds of the Enlightenment* (Manchester: Manchester University Press, 1987), p. 12, on this point generally; Donoghue, 'Imagined More than Women', esp. pp. 200–1, on Britain; and Reis, *Bodies in Doubt*, pp. 17–18, on the United States. In the three countries mentioned, as was sometimes the case in Spain, there was a tendency to consider supposed hermaphrodites as women with oversized clitorises or with a prolapsed vagina. See Guicciardi, 'L'hermaphrodite', pp. 65–7; Graille, *Le troisième sexe*, pp. 61–6; Donoghue, 'Imagined More than Women', p. 201; and, Reis, *Bodies in Doubt*, p. 18.

3. This manuscript is in Spain's Biblioteca Nacional (MS 12966). It was first studied by de la Pascua Sánchez, '¿Hombres vueltos al revés?'

4. 'But are there true hermaphrodites? This question could well be posed in times of igno-rance but should not be asked in enlightened times ... hermaphroditism is a chimera and of those examples of married hermaphrodites either of whom have children, each of them as a man and as a woman, these are puerile fables, brought about by ignorance and through respect for the marvellous, much of which is still to be discounted' (de Jaucourt, 'Hermaphrodite', fols H27–H28, our translation).

5. On this three-dimensional structure and the concept of *mirabilis*, see above, p. 121 n. 59. Original sources in: R. P. A. J. Rodríguez, 'Dissertación II. Sobre la imposibilidad de generación ni comercio por el Demonio íncubo', in *Nuevo Aspecto de Theologia Médico-Moral y ambos derechos o Paradoxas phísico-teológicas-legales*, vol. 2 (Madrid: Imprenta Real de la Gaceta, 1753), pp. 200–15; F. Valderrama, 'Si la muger que pare un Monstruo especie de Bruto, se deba presumir Reo de feo crimen por el Magistrado y como proced-erá contra ella', *Memorias Académicas de la Real Sociedad de Medicina y demás Ciencias de Sevilla*, 5 (1790), pp. 108–20; M. R. P. M. Lorenzo Zambrano, 'Si es posible el con-curso carnal del demonio con criatura humana y en este caso habiendo prole, si es capaz de bautismo', *Memorias Académicas de la Real Sociedad de Medicina y demás Ciencias de Sevilla*, 9 (1790), pp. 409–22. We must not forget that the last person to be burned by the Inquisition in 1781 was a woman accused of fornication with the devil (mentioned by Salamanca Ballesteros, *Monstruos*, p. 204). On the long-standing belief of diabolical intervention in dreams and in the imagination of pregnant women as a cause of monstros-ity, see P. G. Boucé, 'Imagination, Pregnant Women and Monsters in Eighteenth-Century England and France', in G. S. Rousseau and R. Porter (eds), *Sexual Underworlds of the Enlightenment* (Manchester: Manchester University Press, 1987), pp. 86–100. 'Diabol-ical possession' continued to be a category in the medico-legal treatise by J. J. Plenck, *Elementa Medicinae et Chirugiae Forensis* (Madrid: Michaelis Burgos, 1825), pp. 120–1. This was used in the surgeons' colleges in Spain, particularly in San Carlos, Madrid. This college included legal medicine in its curriculum, and this text was translated in 1796 (Granjel, *La Medicina Española del siglo XVIII*, p. 135, and J. Martínez Pérez, 'Sexualidad y Orden Social: la visión médica de la España del primer tercio del siglo XIX', *Asclepio*, 42:2 (1990), p. 119–35, on pp. 123–4). On the popularity of news on teratological beings in the eighteenth century, see Granjel, *La Medicina Española*, pp. 153–4.

6. 'Supongo ciertísimamente en lo sustancial la relación del monstruo en la villa de Fernán Caballero' (I grant that the story of the monster in the town of Fernán Caballero is substantially true), Feijóo wrote in 'Reflexiones filosóficas, con ocasión de una criatura humana hallada poco ha en el vientre de una cabra', in *Cartas Eruditas y Curiosas*, vol. 3 (1750; Madrid: Imprenta Real de la Gaceta, 1774), pp. 327–52, on p. 337. Padre Antonio José Rodríguez rejected the argument by Feijóo that fecund coitus was possible between man and beast. See R. P. A. J. Rodríguez, *Carta respuesta a un ilustre Prelado sobre el feto monstruoso hallado poco ha en el vientre de una cabra y reflexiones críticas que ilustran su historia* (Madrid: Imprenta Real de la Gaceta, 1753). On Feijóo's belief in 'fish-men', nereïds, tritons and other fabulous aquatic monsters, see G. Marañón, *Las Ideas Biológicas del Padre Feijóo* (Madrid: Espasa Calpe, 1954), pp. 223–43.

7. B. J. Feijóo, 'Respuesta a la consulta sobre el infante monstruoso de dos cabezas, dos cuel-los, cuatro manos ... que salió a luz en Medina Sidonia el 24 de febrero del año 1736', in *Cartas Eruditas y Curiosas*, vol. 1 (1742; Madrid: Imprenta Real de la Gaceta, 1777), pp. 78–100, on p. 83.

8. A. J. Mayer, *The Persistence of the Old Regime: Europe to the Great War* (New York: Pan-theon Books, 1981).

9. The notion of 'bare life' is taken from Agamben, *Homo Sacer*.
10. Vázquez García, *La invención del racismo*.
11. Despite his belief in what we would now term 'fabulous' creatures, Padre Feijóo, in contrast to what we would see in Rivilla Bonet (Rivilla Bonet, *Desvíos de la Naturaleza*, fols 35v–36r), did not allude to astrological influences, divine punishment or fornication with the incubus as causes of monsters. In addition, as in Martín Martínez and Antonio José Rodríguez (1703–77), he is sceptical on the creation of monsters by means of imaginative force (Granjel, *La Medicina Española*, p. 135). Despite this, the power of the imagination in the engendering of monsters was still referred to at the beginning of the nineteenth century by authors such as Virey, translated quickly into Spanish, and by Hurtado de Mendoza (Salamanca Ballesteros, *Monstruos*, pp. 228–40).
12. The famous anatomist Martín Martínez (1684–1734), a friend of Feijóo and one of the great innovators of Spanish medicine in the 1700s, tried to explain by means of natural rather than miraculous causes the genesis and short life of a baby boy born in Madrid in 1706 whose heart was positioned outside of the thorax. On this clinical case, see M. Martínez, *Observatio Rara de Corde in Monstroso Infantulo ubi obiter et noviter de motu cordis et sangunis agitur* (Madrid: Francisco Rodríguez, 1750), pp. 231–6. His explanation of monstrosity from an animalculist perspective depends entirely on natural causes. See M. Martínez, *Anatomía Completa del Hombre* (1728; Madrid: Imp. de la Viuda de Manuel Fernández, 1764), p. 202.
13. Canguilhem, 'La Monstruosité et le Monstrueux'; P. Tort, *L'Ordre et les Monstres* (Paris: Le Sycomore, 1980); Daston and Park, *Wonders and the Order of Nature*; F. Jacob, *The Logic of Life: A History of Heredity. The Possible and the Actual*, trans. B. E. Spillman (London: Penguin, 1989), pp. 52–66; J. Farley, *Gametes and Spores: Ideas about Sexual Reproduction 1750–1914* (Baltimore, MD: Johns Hopkins University Press, 1982); and M. Hagner, 'Utilidad científica y exhibición pública de monstruosidades en la época de la Ilustración', in A. Lafuente and J. Moscoso (eds), *Monstruos y Seres Imaginarios en la Biblioteca Nacional* (Madrid: Ministerio de Educación y Cultura, Biblioteca Nacional, 2000), pp. 105–28. Hagner also refers to the 'aesthetic' uses of the monster. The collection and exhibition of monsters was current throughout the eighteenth century.
14. Feijóo indicated that the birth of a human baby from a goat made him change his ideas from the 'ovevos u ovuistas' (ovist) school to the animalculist school (Feijóo, 'Reflexiones filosóficas', pp. 344–5).
15. J. Moscoso, 'Monsters as Evidence: the Uses of the Abnormal Body during the Early Eighteenth Century', *Journal of the History of Biology*, 31 (1998), pp. 355–82.
16. See the classic work of H. Daudin, *De Linné a Lamarck. Méthodes de la classification et idée de série en Botanique et en Zoologie (1740–1790)* (Paris: Alcan, 1926–7), and *Cuvier et Lamarck. Les classes zoologiques et l'idée de série animale*, 2 vols (Paris: Alcan, 1926–7).
17. Javier Moscoso undertakes a quantitative analysis of the incidence of observations on monsters appearing in the *Journal de Savants* (twenty-five articles between 1665 and 1710 and just eight between 1710 and 1750) and in *Philosophical Transactions* (forty communicacions between 1665 and 1712 and sixteen between 1775 and 1810). The evolution is similar in the Germanic *Acta eruditorum* (Moscoso, 'Monsters as Evidence', pp. 359–60).
18. Ibid., p. 360. See also Salamanca Ballesteros, *Monstruos*, p. 17, on the eighteenth century as a point of inflection; and especially ch. 9 ('The Enlightenment and the Antimarvelous') in Daston and Park, *Wonders and the Order of Nature*, pp. 329–63.
19. 'Porque entre los autores compiladores de prodigios, hay no pocos fáciles en creer, y ligeros en escribir. Son muchos los hombres que se complacen en referir portentos y rara

vez falta quien eternice con la estampa sus ficciones, como si fuesen realidades' (Because among those authors who have compiled stories on prodigies, there are many who are easily convinced and many who write without substance. There are many men who are content to refer to portents and there are very often those who seal their fictions with the stamp of eternity, as if they were fact) (Feijóo, 'Respuesta a la consulta', p. 80).

20. F. Sánchez Blanco Parody, *Europa y el Pensamiento Español del siglo XVIII* (Madrid: Alianza Universidad, 1991), pp. 134–72.

21. On the important proliferation of volumes on hermaphroditism in Europe in the eighteenth century (medical texts, travel literature, novels with hermaphrodite protagonists), see Vázquez García and Moreno Mengíbar, *Sexo y Razón*, pp. 199–200.

22. For Martín Martínez, masculinized women are in reality women with large clitorises: 'en el fervor de acto venéreo [el clítoris] se hincha y enfurece como el miembro viril; y en algunas ha crecido tanto, que han podido abusar de la Venus con otras mugeres, y dar ocasión al vulgo para creer las fábulas de hembras convertidas en varones, ansí como a las de hombres transformados en mugeres, ha dado motivo el ocultarse del todo el pene' (in the heat of the venereal act [the clitoris] swells and becomes inflamed like the male member; and in some women it has grown so much that they have been able to abuse Venus with other women, allowing the uninformed to believe fables of women turned into men, as well as men turned into women, their penis disappearing entirely) (Martínez, *Anatomía Completa*, p. 188). This work has been viewed as the best Spanish morphological work of the first half of the eighteenth century by López Piñero et al., 'Martínez, Martín', in *Diccionario Histórico de la Ciencia Moderna en España*, vol. 2, p. 34.

23. 'De la mutación de sexos en una misma persona no discurro, porque repugna totalmente al orden y leyes de la naturaleza; y cualquiera a la menor reflexión la conoce imposible' (Of the change of sex in one person I cannot accept because this goes against order and nature; anyone with the least reflection would declare it impossible); L. Hervás y Panduro, *Historia de la Vida del Hombre o idea del Universo*, vol. 1 (Madrid: Aznar, 1789), p. 189.

24. Antonio Jacobo del Barco y Gasca, Vicar of Huelva from 1747, was an enlightened figure who applied the same kind of critical approach as Feijóo, refuting supposed sex changes. See A. J. del Barco y Gasca, 'Examen crítico de una rara transmutación de sexos en persona del femenino', in *Cartas Familiares, Varias y Curiosas, dispuestas para honesta diversión* (manuscript), vol. 3, letter 29 (Seville: Biblioteca de la Real Academia de Buenas Letras, 1770–1), fols 199–221. On Barco y Gasca and his intervention in this case, see S. Muñoz-Prián, 'Identidades transgenéricas en la España del Antiguo Régimen. Un caso de cambio de sexo en la Andalucía del siglo XVIII' (unpublished MA thesis, Universidad de Cádiz, 2009).

25. 'Aquí pertenece refutar las "historietas" que se refieren a la alteración o cambios de los sexos; la doctrina expuesta sobre las causas de la nymphomania y del hermafroditismo, son las que han hecho se crean estas apariencias' (Here it is pertinent to refute those 'stories' that refer to the alteration or change of sex. Doctrine on the causes of nymphomania and hermaphroditism has led to these beliefs); J. Fernández del Valle, *Cirugía forense, general y particular*, vol. 3 (Madrid: Imprenta de Aznar, 1797), p. 18.

26. The 'hermaphrodite' entry in the *Diccionario de Autoridades* (1732) appears not to doubt the existence of these beings, although, in contrast to the *Tesoro* by Covarrubias, it considers them to be monstrous: 'Hermaphrodita. La persona que tiene los dos sexos de hombre y muger, que por otro nombre se llama Andrógeno. Tienen los autores varias opiniones del motivo o causa de esta monstruosidad y por extensión se dice de otras cosas' (Hermaphrodite. The person that possesses both sexes male and female and who

goes by the other name of Androgyne. Authors have various opinions on the motive or cause of this monstrosity); Various Authors, *Diccionario de la Lengua Castellana en que se explica el verdadero sentido de las voces, su naturaleza y calidad*, vol. 3 (1732; Madrid: Ediciones Turner, 1977), p. 144.

27. Feijóo 'Respuesta a la consulta', p. 83.
28. Martínez, *Anatomía Completa*, p. 202.
29. Ibid., p. 182.
30. Hervás y Panduro, *Historia de la Vida del Hombre*, p. 185.
31. Ibid., p. 184.
32. Ibid., p. 185.
33. Ibid., pp. 186–8.
34. Ibid., p. 188.
35. Ibid., p. 189.
36. The thesis on hermaphrodites appears in the first chapter under the title 'Reasons Against the Existence of an Hermaphroditical Nature in Human Bodies', in Parsons, *Mechanical and Critical Inquiry*, pp. 1–37.
37. A. Sáiz Carrero, 'Real Colegio de Cirugía de San Carlos', *Urología Integrada y de Investigación*, 14:2 (2009), pp. 188–206.
38. The Royal College of Barcelona was where the need to include legal surgery in the curriculum was defended most ardently. See J. Martínez Pérez, 'Colegios de Cirugía y Medicina Legal: una expresión de los procesos de intercambio entre fuerzas armadas y sociedad a finales del siglo XVIII', in E. Balaguer and E. Giménez (eds), *Ejército, Ciencia y Sociedad en la España del Antiguo Régimen* (Alicante: Instituto de Cultura Juan Gil-Albert, 1995), pp. 493–511, on p. 505.
39. Ibid., pp. 508–11.
40. Sáiz Carrero, 'Real Colegio de Cirugía', p. 8.
41. *Gaceta de Madrid*, 21 January 1803, p. 475.
42. J. Fernández del Valle, *Cirugía forense, general y particular*, vol. 1 (Madrid: Imprenta de Aznar, 1796), pp. 256–7.
43. Ibid., p. 257.
44. Ibid.,. 260.
45. Donoghue, 'Imagined More than Women'.
46. Fernández del Valle, *Cirugía forense*, vol. 1, pp. 260–1.
47. Ibid., pp. 261–2.
48. Fernández del Valle, *Cirugía forense*, vol. 3, p. 295.
49. Ibid., p. 18.
50. J. Martínez Pérez, 'Medicina, Liberalismo y Legislación: Ramón López Mateos (1771–1814) y sus *Pensamientos sobre la razón de las leyes*', *Asclepio. Revista de Historia de la Medicina y de la Ciencia*, 40:2 (1988), pp. 209–46.
51. Fernández del Valle, *Cirugía forense*, vol. 1, p. 262.
52. The same cannot be said for the work's scope. For Zacchia the legal doctor is simply an aid to the judge in certain questions such as witchcraft, poisoning, violent attacks, births and ecclesiastical disputes. For Plenck, trained in the German cameralist tradition, legal medicine is a branch of the art of government, which contributes to the increase in population and the maintenance of the quality of public health. See the section on the 'political' aspects of medicine in the *Elementa Medicinae*, the Latin text of which was presented by Dr A. Vallejo and published in Madrid by Michaelis Burgos in 1825 (pp. 120–5). On the differences between pre-modern and modern legal medicine, see

Martínez Pérez, 'Sexualidad y Orden Social', esp. pp. 121–3, and J. L. Peset and M. Peset, 'Estudio preliminar', in *Lombroso y la Escuela Positivista Italiana* (Madrid: CSIC, 1975), pp. 80–1.

53. See Marchetti, *L'invenzione della bisessualità*, pp. 65–110.

54. Plenck, *Elementa Medicinae*, p. 116.

55. Ibid., pp. 117–18.

56. Ibid., pp. 118–19.

57. Ibid., p. 119.

58. 'De Androgynis autem veris utrumque esse possibile' (ibid., p. 120).

59. 'Atrocem et iniquissimam fuisse veterum legem, auqe homines dubii sexus quos ipsa natura jam severius tractavit, cum morte puniebat' (ibid.).

60. Cases which also demonstrate this transition include Mariana (Córdoba, 1774), Martina Parra (Nueva Granada, Colombia, 1803) and Antonio Martínez (Chiclana, 1813). The first of these, a nun from the Augustine convent in Cordoba, relates how after two years in the convent she was expelled as a 'machihembra' (man-woman). Subsequently, she married Francisco Gómez Linares and went to live in Montilla. Once she was widowed, she intended to marry her former husband's cousin, 'haciendo papel de varón' (taking the role of the man). Barco y Gasca, who was consulted about the case, admitted that this was a case of hermaphroditism and applied canon law by which the elected sex could not be altered. He also argued that the perfect hermaphrodite, whether male or female, was naturally impossible. The hermaphrodite could not reproduce and for this reason should not enter into marriage. Otherwise, unreproductive relations would be knowingly approved and this was against natural law as mollities and sodomy. The report by Barco y Gasca has been transcribed by Muñoz-Prián, *Identidades*. The case of Martina Parra is founded on an accusation by the Real Audiencia de Santafé that she was a hermaphrodite who cohabited with her employer, Ana María Martínez, for whom she worked as a housemaid. According to the statement made by Martínez, Parra possessed a member 'como de hombre' (like that of a man). The doctors who examined Parra declared that she was a woman of normal characteristics. They suggested that the basis of the 'fable' of 'hembras convertidas en varones' (women becoming men) was the size of the clitoris. However, they never place hermaphroditism as such in doubt. See on this point C. Giraldo, 'Una hija de Hermes y Afrodita en un expediente de 1803' (2008), at http://www.incertidumbre.org/index.php?option=com_content&task=view&id=85 &Itemid=55 [accessed 15 May 2013]. The case of the seaman Antonio Martínez comes from the beginning of the nineteenth century. This individual, of nineteen years, was baptized as a woman but worked as a man. He was examined by the medical authorities in Cuba as he argued that he was a hermaphrodite and therefore could not serve in the navy. Dr Romay, who published a brief account of the case in 1813 (*Diario del Gobierno de La Habana*, May 1813), admitted that this was a case of hermaphroditism, although he pointed out that in the higher species these individuals were incapable of inseminating or conceiving. The doctor also referred to the case of Fernanda Fernández; Tomás Romay, cited in P. Marqués de Armas, 'El Monstruo Humano' (2002), at http://www.habanaelegante.com/Summer2002/Panoptico.html [accessed 15 February 2013].

61. D. Vidal, *Cirugía Forense o arte de hacer las relaciones chirurgico-legales* (1783; Zaragoza: Imprenta de las Heras, 1814); facsimile edition by J. Corbella (Barcelona: Seminari Pere Mata, Universidad de Barcelona, 1987).

62. V. Mitjavila, *Compendio de Policía Médica*, facs. edn by J. M. Calbet and J. Corbella (1803; Barcelona: Universidad de Barcelona, 1983).

63. On this innovatory institution, see A. Zarzoso Orellana, 'La Pràctica Mèdica a la Cata- lunya del segle XVIII' (unpublished doctoral thesis, Universidad de Barcelona, 2003), pp. 129–94.

64. M. Foucault, *Securité, Territoire, Population. Cours au Collège de France 1977–1978* (Paris: Gallimard-Le Seuil, 2004). On the development of police science in Spain, see P. Fraile, *La Otra Ciudad del Rey. Ciencia de la Policía y organización urbana en España* (Madrid: Celeste Ediciones, 1997); A. Zarzoso Orellana, 'Policía y Ciencia de la Policía en el Discurso Urbanístico a Finales del Antiguo Régimen', *Asclepio*, 50:1 (1998), pp. 259–64; and Vázquez García, *La invención del racismo*.

65. On the importance of cameralism in Spanish Enlightened thought, see E. Lluch, 'El Cameralismo en España', in E. Fuentes Quintana (ed.), *Economía y Economistas Espa- ñoles. Tomo III. La Ilustración* (Barcelona: Galaxia Gutenberg, 2000), pp. 721–8.

66. On the rise of public health and health policy initiatives in Europe from 1750 to 1830, see G. Rosen, *A History of Public Health* (1958; Baltimore, MD and London: Johns Hopkins University Press, 1993).

67. E. Rodríguez Ocaña, 'El resguardo de la salud. Administración sanitaria española en el siglo XVIII', in *Salud Pública en España. Ciencia, Profesión y Política, siglos XVIII–XIX* (Granada: Universidad de Granada, 2005), pp. 17–48.

68. Vázquez García, *La invención del racismo*.

69. Martínez Pérez, 'Sexualidad y Orden Social', p. 123.

70. 'Los objetos de la Cirugía forense se pueden reducir a dos, uno próximo y otro remoto: el primero se dirige a saber y conocer la verdad; el segundo es consiguiente y conspira a conservar la buena armonía y tranquilidad de un Estado' (The objectives of forensic surgery can be reduced to two, one immediate and the other more distant: the first is to know truth; the second is derived from this and seeks to conserve the true harmony and tranquility of a State) (Fernández del Valle, *Cirugía forense*, vol. 1, p. 62).

71. On the 'police state' in this older sense, cf. Foucault, *Securité, Territoire, Population*, pp. 341–70, and M. Dean, *Governmentality: Power and Rule in Modern Society* (London: Sage, 1999), pp. 89–96.

72. On the importance of gender differences in German cameralism, see I. V. Hull, *Sexuality, State and Civil Society in Germany, 1700–1815* (Ithaca, NY: Cornell University Press, 1996), pp. 294–8.

73. On the role of the theologians Carpzov and Dedekener at the beginning of the seven- teenth century, see Marchetti, *L'invenzione della bisessualità*, p. 336.

74. This law specifically on hermaphrodites allowed parents alone to decide in cases of doubt- ful sex. This decision could be altered by the subject him/herself on reaching eighteen years of age, but if the rights of a third person were affected, a medical-legal diagnosis could be requested. This effectively trumped the parents' and the individual's decision on the sex to be adopted. See L. Thoinot, *Tratado de Medicina Legal, traducido, anotado y adicionado con referencia a la legislación española y americana por W. Coroleu, Secretario Perpetuo de la Real Academia de Medicina de Barcelona*, vol. 2 (Barcelona: Salvat Editores, 1928), p. 109.

75. F. E. Foderé, *Las leyes ilustradas por las Ciencias Físicas o Tratado de Medicina Legal y de Higiene Pública*, 8 vols (1798; Madrid: Imp. de la Administración del Real Arbitrio de Beneficencia, 1801–3), vol. 1, pp. 1–2: 'y como las leyes no pueden ser buenas si no están de acuerdo con el hombre, con su corazón, necesidades, clima y género de vida a que están sujetos los diferentes pueblos, deben los legisladores y los magistrados consultar la medicina, vasto código de las leyes de la física animal, antes de pensar en establecer nue- vas instituciones o para darlas todo el grado de utilidad que son capaces de recibir' (and as

laws cannot be good if they are not made in accordance with man, with his heart, needs, climate and way of life of different peoples, legislators and magistrates should consult medicine, a vast code of the laws of animal physique, before establishing new institutions or bestowing on them the highest degree of utility they are capable of receiving).

76. In chronological order: P. M. Peiró and J. Rodrigo, *Elementos de Medicina y Cirugía Legal arreglados a la Legislación Española* (Zaragoza: Imprenta de Mariano Peiró, 1832) (new editions 1839, 1841 and 1844); P. Mata, *Vademecum de Medicina y Cirugía Legal*, vol. 1 (Madrid: Imprenta Calle de Padilla, 1844); P. Mata, *Tratado de Medicina y Cirugía Legal* (Madrid: Carlos Bailly Baillière, 1846) (republished several times); R. Ferrer y Garcés, *Tratado de Medicina Legal* (Barcelona: Imprenta de P. Riera, 1847); and A. Rossell, *Manual de Medicina Legal* (Madrid: Ramón Rodríguez Rivera, 1848). The translation of the *Traité de Médecine Légale* by Mateo Orfila was completed in 1847 from the 1835 edition (the *Traité* had been published for the first time in 1821 with the title *Leçons faisant partie du cours de Médecine Légale*). On the context of French legal medicine around the time of Orfila, see R. Huertas, *Orfila, Saber y Poder Médico* (Madrid: CSIC, 1988), pp. 31–3. In 1843 the Chair in Legal Medicine was established at the University of Madrid, and in 1845 that of the University of Barcelona. In 1853 the first Spanish journal specializing in legal medicine was created with the title *Repertorio de Higiene Pública y Medicina Legal*, edited by Manuel Álvarez Chamorro. In 1855 the Cuerpo Provisional Médico Forense of Madrid was established (Juan Querejazu Hartzenbusch, the translator of Tardieu in Spain, was a member), and 1862 saw the organization of the medical legal profession nationally. For these details, see M. Pérez de Petinto and M. Bertomeu, 'Comienzo y actualidad (en 1951) de la trayectoria corporativa médico-forense', *Revista Española de Medicina Legal*, 23 (1999), pp. 5–43, esp. pp. 6–10.

77. 'Las leyes entienden en arreglar la moralidad de las acciones; y la medicina en averiguar los instrumentos que la determinan y modifican. Sin un exacto discernimiento de la variedad de circunstancias que pueden concurrir a determinar y modificar esta moralidad, sugerido por la ciencia de la vida y de la muerte, mal podrá el legislador ajustar como debe sus preceptos a las insinuaciones de la naturaleza' (Laws are designed to correct the morality of actions; medicine to determine the instruments which conform and modify morality. Without a precise identification of the variety of circumstances that may converge to determine and modify this morality, as proposed by the science of life and death, it will be a difficult task for the legislator to adjust as he should his precepts to the insinuations of nature); R. López Mateos, *Pensamientos sobre la razón de las leyes*, cited in Martínez Pérez, 'Sexualidad y Orden Social', p. 126. On 'classical' liberal governmentality, see Dean, *Governmentality*, pp. 113–30, and Vázquez García, *La invención del racismo*.

78. 'Todo hombre, generalmente hablando, en habiendo llegado a la pubertad, siente en su interior un poderoso estímulo que le incita a la propagación de su especie; pero tanto como una unión desarreglada e ilegítima no conviene al Estado, se debe favorecer, quanto sea posible, la conyugal, con atención a que tiene cuenta a todo gobierno que sus Reynos y Provincias estén competentemente poblados; y supuesto que las ventajas y prosperidad de una población están en razón directa de la robustez y sanidad de sus moradores, proporcionadas a la naturaleza del suelo en que viven' (All men, in general, on arriving at puberty, experience in their insides a powerful stimulus that incites the propagation of the species. As much as an illegitimate and careless union is inconvenient for the State, as far as is possible, the conjugal union should be favoured, and all governments should ensure that their Kingdoms and Provinces are sufficiently populated. The advantages and prosperity of a population are in direct relation to the robustness and health of its inhab-

itants, in accordance with the nature of the soil where they live) (Mitjavila, *Compendio de Policía Médica*, p. 69).

79. On hermaphroditism and the question of 'same-sex marriages', see A. D. Dreger, *Hermaphrodites and the Medical Invention of Sex* (Cambridge, MA: Harvard University Press, 1998), pp. 119–26.

80. See I. Morant and M. Bolufer Peruga, *Amor, Matrimonio y Familia. La construcción histórica de la familia moderna* (Madrid: Síntesis, 1998), and M. Bolufer Peruga, 'Lo Íntimo, lo Doméstico y lo Público. Representaciones y Estilos de Vida en la España Ilustrada', *Studia Histórica, Historia Moderna*, 19 (1998), pp. 85–116, esp. pp. 109–10.

81. G. M. Jovellanos, 'Sátira primera a Arnesto', in *Obras Completas, vol. I: Obras literarias* (1786; Oviedo: Centro de Estudios del Siglo XVIII, Ilustre Ayuntamiento de Gijón, 1984), pp. 220–7, on pp. 221–5, reproduces the connections between luxury and luxuriousness.

82. Ibid., p. 225, contrasts this commerce between beautiful women and older rich men with the old custom whereby the most beautiful women were handed over as prizes for military valour. See also ibid., p. 227 n. 15, by the editor José Miguel Caso González.

83. Ibid., p. 224.

84. R. Alcalá Flecha, *Literatura e Ideología en el Arte de Goya* (Zaragoza: Diputación General de Aragón, 1988), pp. 349–51.

85. 'For a long time, the individual was vouched for by the reference of others and the demonstration of his ties to the commonweal (family, allegiance, protection); then he was authenticated by the discourse of truth he was able or obliged to pronounce concerning himself' (Foucault, *The History of Sexuality, Volume 1*, p. 58).

86. Cleminson and Vázquez García, 'The Hermaphrodite, Fecundity and Military Efficiency'.

87. F. E. Foderé, *Les lois éclairées par les sciences physiques, ou traité de médecine légale et d'hygiène publique* (Paris: chez Croullebois et chez Deterville, 1813), pp. 355–66, rejects the existence of actual cases of 'androgyny', although he does recognize the 'monstrosities' that can give rise to such errors of diagnosis. He believes that physical and visual examination will dispell any uncertainities; in this way, marriages can be annulled as a result of 'impotency'. He also indicates (Foderé, *Les lois éclairées*, pp. 178–9) that the sex of a newborn can be the subject for debate, but he attributes any doubts to a lack of precision on behalf of 'sages-femmes illitérées' (illiterate midwives). R. López Mateos, *Pensamientos sobre la razón de las leyes* (Madrid: Imp. Gómez Fontenebro y Compañía, 1810), pp. 111–21, refers to impotence and sterility but does not mention hermaphroditism.

88. See, for example, a case in Cadiz in 1717; the bishop instructed the grandmother of the person in question to watch over the couple for three nights to see if the marriage was consummated. See A. Morgado García, 'El divorcio en el Cádiz del siglo XVIII', *Trocadero. Revista de Historia Moderna y Contemporánea*, 6–7 (1994–5), pp. 125–37, esp. pp. 133–4.

89. The classic study by P. Darmon, *Trial by Impotence* (London: Chatto & Windus, 1985), pp. 210–17, suggests that from the time of the Napoleonic Code and through most of the nineteenth century, the category of impotence disappeared as a motivation behind divorce and became subsumed into the category of 'identity error'. However, in canon law, eighteenth-century strictures were maintained.

90. C. J. Esdaile, *The Spanish Army in the Peninsular War* (Manchester and New York: Manchester University Press, 1988), p. 10.

91. On the requirements and exemptions detailed in the 1842 set of rules and its association with the establishment of the 'quintas' recruiting system in 1830, see Cleminson and Vázquez García, 'The Hermaphrodite, Fecundity and Military Efficiency', p. 78.

92. G. Bravo de Sobremonte, *Operum Medicinalium*, four vols (Lyon: L. Arnaud, 1671); García Carrero, *Disputationes Medicae*; Martín del Río, *La Magia Demoníaca*.

93. These two authors opened up Spanish thought to Enlightenment currents: 'Macanaz, Martín Martínez y Feijóo, cada uno en su campo limitado de actividad, abren el camino a quienes ya en la segunda mitad de la centuria se esfuerzan por incorporar a España al movimiento cultural europeo' (Macanaz, Martín Martínez and Feijóo, each within his own area of activity, opened the way to those in the second half of the century who tried to harness Spain to European cultural trends) (Granjel, 'El Pensamiento Médico de Martín Martínez', in *Médicos Españoles*, p. 171). On the construction of the female body in Spain in the eighteenth century, the work of Mónica Bolufer Peruga is essential. For a synthesis and overview of the debates, see her *Mujer e Ilustración. La construcción de la feminidad en la España del siglo XVIII* (Valencia: Institució Alfons el Magnànim, 1998), and 'Cos femení, cos social. Apunts d'historiografia sobre els sabers mèdics i la construcció cultural d'identitats sexuades (segles XVI–XIX)', *Afers. Full de recerca i pensament*, 33–4 (1999), pp. 531–50.

94. B. J. Feijóo, 'La Doctrina Hipocrática no debe tomarse por norma de Medicina', in *Teatro Crítico Universal*, vol. 8 (1739; Madrid: Imp. de Don Joaquín Ibarra, 1779), pp. 328–39. On this question, see Marañón, *Las Ideas Biológicas*, pp. 216–17.

95. Feijóo, 'La Doctrina Hipocrática', p. 333.

96. B. J. Feijóo, 'Defensa de las Mujeres', in *Teatro Crítico Universal*, vol. 1 (1726; Madrid: Imp. de Don Joaquín Ibarra, 1778), pp. 325–98, on p. 330. On the huge impact of this work upon the intellectual currents of Spain in the eighteenth century, consult Bolufer Peruga, *Mujer e Ilustración*, pp. 28–59.

97. Feijóo, 'Defensa de las Mujeres', p. 331; Eiximenis, *Carro de las Donas*.

98. A similar but less marked understanding of physical differences appears in J. Amar y Borbón, *Discurso sobre la Educación Física y Moral de las Mujeres* (1790; Madrid: Cátedra, 1994), p. 63.

99. M. Bolufer Peruga, 'Galerías de "Mujeres Ilustres" o el sinuoso camino de la excepción a la norma cotidiana (ss. XV–XVIII)', *Hispania*, 60:204 (2000), pp. 181–224.

100. Feijóo, 'Defensa de las Mujeres', pp. 359–60.

101. Ibid., p. 389.

102. M. Onaindía, *La construcción de la nación española. Republicanismo y nacionalismo en la Ilustración* (Barcelona: Ediciones B, 2002), pp. 37–8.

103. Ibid., p. 90.

104. On various figures such as Sobre Claudious, Tarquin, Almanzor and Rachel the Jewess as monsters or despots, see ibid., pp. 100, 119–20, 182 and 216. The same author argues that the motif of 'monstrous governments' coming from Montesquieu gave rise to the notion of Napoleon as a monster in the War of Independence. See ibid., pp. 236 and 317. A genealogy of the 'political monster' can be found in Foucault, *Les Anormaux*, pp. 87–97.

105. Onaindía, *La construcción de la nación española*, p. 313.

106. Ibid., p. 333, on 'inconstant' Africans.

107. This argument did not eliminate, as we shall see in G. M. Jovellanos, 'Sátira segunda a Arnesto. Sobre la mala educación de la nobleza', in *Obras Completas, vol. I: Obras literarias* (1787; Oviedo: Centro de Estudios del Siglo XVIII, Ilustre Ayuntamiento de Gijón, 1984), pp. 227–44, the old critique of the nobility as effeminate. The stereotype

continues in Feijóo (Cartagena-Calderón, "'Él es tan rara persona'", esp. pp. 156–7), in *El señorito mimado* by Tomás de Iririarte (Onaindía, *La construcción de la nación española*, p. 210) and in the historical essays by Jovellanos, who refers to the decadent Visigoth nobility whose betrayal, under the reign of Don Rodrigo, facilitated the Arab invasion (ibid., p. 245).

108. Jovellanos, 'Sátira segunda a Arnesto', pp. 228–35.

109. Ibid., pp. 228–33.

110. Ibid., p. 233.

111. Ibid., pp. 233–4.

112. Ibid., p. 234.

113. As J. de Cadalso, *Cartas Marruecas* (1789; Madrid: Imprenta de Sancha, 1793), p. 164, wrote: 'la abundancia ha producido el luxo, a este luxo se ha seguido la afeminación, de esta afeminación ha nacido la flaqueza, de la flaqueza ha dimanado su ruina' (abundance has brought luxury, to which has been added effeminacy, which brought weakness, which has in turn brought about its [the nation's] ruin).

114. Onaindía, *La construcción de la nación española*, pp. 227–30.

115. Cadalso, *Cartas Marruecas*, p. 10.

116. Ibid., pp. 13–14: 'la mezcla de las naciones en Europa, ha hecho admitir generalmente los vicios de cada una, y desterrar las virtudes respectivas. De aquí nacerá, si ya no ha nacido, que los nobles de todos los países tengan igual despego de su patria' (the mixture of nations in Europe has in general exposed the vices of each and has eliminated their respective virtues. From this will result, if this is not already so, an equal disregard of each nation's nobility for their country).

117. On their friendship, see Granjel, 'El Pensamiento Médico de Martín Martínez', pp. 193–5, and Marañón, *Las Ideas Biológicas*, pp. 118–24.

118. 'El tercer género de partes contenidas en el vientre inferior, son las que sirven a la generación, y de éstas unas son comunes a ambos sexos, como los vasos espermáticos, testículos y vasos deferentes, y otras propias de cada sexo, como en los varones la epídidimis, vesículas seminales y miembro viril, y en las mugeres el útero. Estas partes son nobilísimas y principales en orden a la especie, y fueron dadas por la naturaleza, para que ya que los individuos no pueden perpetuarse, se perpetúe y no se envejezca la especie, renovada en cada individuo' (The third kind of parts contained in the lower abdomen are those that serve generation and of these some are common to both sexes, such as the sperm ducts, testicles and vas deferens, and others are possessed by one sex, as the epididymis, seminal tracts and virile member in males, and in women the uterus. These parts are noble and integral to the species and were given by nature, so that since individuals cannot procreate by themselves, the species is reproduced and does not die off, being renovated in each new individual) (Martínez, *Anatomía Completa*, p. 159).

119. On the difference between the spermatic ducts and the female ovaries with respect to their male counterparts, see ibid., p. 178.

120. On the form of the womb as convenient for the expulsion of the foetus, see ibid., p. 182; on the role of the 'ligamentos redondos' (round ligaments) facilitating birth, see ibid., p. 183; on the uterus as 'fecundo campo de la generación' (fecund site of generation), see ibid., p. 184; on the function of labia in opening the vulva during birth, see ibid., p.187. Regarding the muscles of the clitoris, Martínez states the following: 'parece que sirven de cerrar el orificio de la vulva y comprimir en el coito el pene, y no de elevar el clítoris o arrojar el esperma, como otros presumen' (they appear to have the function of closing the orifice of the vulva and of compressing the penis during coition and not of elevating

the clitoris or ejaculating sperm, as some suppose) (ibid., p. 183). In this way, the lack of similarity between the penis and the clitoris is emphasized. Finally, the vagina possesses a sphincter that prevents the entry of air and avoids 'enfriar el esperma espirituoso masculino antes que penetre por las tubas a los ovarios' (the cooling down of the spirited sperm of the male before it penetrates the tubes of the ovaries) (ibid., p. 192).

121. Ibid., p. 187.

122. Ibid., pp. 188 and 192.

123. On the construction of nymphomania in France, see J. M. Goulemot, 'Fureurs utérines', *Dix-Huitième Siècle*, 12 (1980), pp. 97–111, and Y. Knibiehler and C. Fouqué, *La Femme et les Médecins* (Paris: Hachette, 1983), pp. 144–8. For Britain, see G. S. Rousseau, 'Nymphomania, Bienville and the Rise of Erotic Sensibility', in P. G. Boucé (ed.), *Sexuality in Eighteenth-Century Britain* (Manchester: Manchester University Press, 1982), pp. 95–119. For Spain, see F. Vázquez García, 'Ninfomanía y construcción simbólica de la femineidad (España, siglos XVIII–XIX)', in C. Canterla (ed.), *VII Encuentro de la Ilustración al Romanticismo. La Mujer en los siglos XVIII y XIX* (Cadiz: Pub. Universidad de Cádiz, 1994), pp. 125–135.

124. Ventura Pastor, whose texts have been considerd to be 'el mejor testimonio del desarrollo logrado por la Tocología ilustrada' (the best example of the development attained by illustrated Tocology) in Spain (Granjel, *La Medicina Española en el siglo XVIII*, pp. 222–3), describes 'furor uterino' (uterine fury) as an 'impúdica enfermedad' (shameless illness). Mentioning Baudelocq, Ventura cites, perhaps for the first time in Spain, the operation of clitoridectomy as a means of curing this illness: 'otras veces (dice [Baudelocq]) ha sido preciso separarle [al clítoris] de las mujeres jóvenes a causa de hallarse consumidas del marasmo y próximas a quedarse abatidas y enteramente extenuadas con motivo de las copiosas evaquaciones uterinas de todas clases suscitadas por la irritación mecánica y continua de esta parte' (on other occasions ([Baudelocq] states) it has been necessary to remove [the clitoris] from young women because they have been consumed by miasmas and have been defeated and entirely exhausted by copious uterine evacuation of all kinds brought about by the continuous mechanical irritation of this part); J. Ventura Pastor, *Preceptos Generales sobre las Operaciones de los Partos* (Madrid: Joseph Herrera, 1789), p. 35.

125. Bolufer Peruga, *Mujer e Ilustración*, p. 69.

126. Foderé, *Las leyes ilustradas*, vol. 1, pp. 48–51.

127. 'Lo más débil y sensible de la muger la inutilizó para grandes fatigas, y para negocios de discusión seria y detenida; al paso que la proporcionó a impresiones las más ligeras, y a que tomase interés en cosas despreciables o de poca importancia. La conformación particular de los huesos de las caderas y demás que conforman la pelvis facilitaba la postura sentada, como también lo más abultado de sus músculos por su gran texido celular, y mayor diámetro de su base, haciéndola declinar a ocupaciones sedentarias y tranquilas' (Woman's weakness and sensitivity have precluded her from great exertion and from serious and prolongued discussion, at the same time allowing her to perceive more delicate impressions and to interest herself in useless or unimportant things. The particular structure of her hip bones and other bones that make up the pelvis facilitate a seated posture, as does the bulk of her muscles due to the larger cellular texture, the larger diameter of her posterior, making her inclined towards sedentary and tranquil occupations); López Mateos, *Pensamientos sobre la razón de las leyes*, cited in Martínez Pérez, 'Sexualidad y Orden Social', pp. 128–9.

128. M. Bolufer Peruga, 'Literatura encarnada: modelos de corporalidad femenina en la Edad Moderna', in S. Mettalía and N. Girona (eds), *Aún y más allá: mujeres y discursos* (Caracas: Ex Cultura, 2002), pp. 205–15, esp. pp. 209–10.

129. J. L. Peset, *Ciencia y Marginación. Sobre negros, locos y criminales* (Barcelona: Crítica, 1983), p. 9; M. Foucault, *'Il Faut Défendre la Société'. Cours au Collège de France, 1976* (Paris: Seuil, 1997), pp. 70–3.

130. This meant that hermaphroditism was excluded from the higher orders of Nature, and in these species sexual inclination was more acute: 'el hermafroditismo era menos aplicable a las especies que, poseyendo sentidos y membranas, podían más fácilmente moverse y conocer sus semejantes: también la naturaleza ha separado los sexos en los animales que se transportan con facilidad y están provistos de sentidos. Pero para obligar a los sexos a que se buscasen, ha sido necesario darles el sentimiento del gozo más vivo y delicado que a los hermafroditas. Estos, al contrario, debían tener deseos más moderados y limitados para no destruirse a sí mismos con solicitaciones continuas de amor. ¿Qué abuso, que pronta muerte no se seguiría al hermafrodismo completo en seres tan ardientes en amor como las aves, los cuadrúpedos y el hombre? Este estado no conviene sino a las especies frías y poso sensibles, como los animales imperfectos y las plantas' (hermaphroditism was less applicable to those species that, possessing sense and membranes, could move easily and meet those similar to them. Nature has also distinguished the sexes in animals that can move easily and which have sense. But in order to oblige the sexes to look for one another it has been necessary to make them more sensitive to a greater pleasure than that experienced by hermaphrodites. Hermaphrodites, on the other hand, must have more moderate and limited desires in order not to destroy themselves through continuous demands for amorous activity. What degradation, what quick death would await complete hermaphroditism in species so ardent in lovemaking as birds, quadrupeds and man? Such a state is only convenient for cold species and those with little sensation such as imperfect animals and plants); J. J. Virey, *Tratado Histórico y Fisiológico Completo sobre la Generación, El Hombre y la Muger* (Madrid: Imprenta de Antonio Martínez, 1821), pp. 24–5.

131. '¡Cuántas precauciones y cuánta prudencia necesita el médico para dirigir la salud de una organización tan frágil y movible como es la de la muger en todos los estados de su vida!' (How many precautions and how much prudence must the doctor display in order to direct the health of so fragile a being as woman throughout all the phases of her life!) (ibid., p. 155).

132. J. M. J. Vigarous, *Curso Elemental de las Enfermedades de las Mugeres*, 2 vols (Madrid: Imp. de Juan de Brugada, 1807).

133. B. de Viguera, *La Fisiología y Patología de la Muger o sea historia analítica de su constitución física y moral, de sus atribuicones y fenómenos sexuales y de todas sus enfermedades*, 2 vols (Madrid: Imprenta de Ortega y Compañía, 1827).

134. P. Roussel, *Sistema Físico y Moral de la Muger* (Madrid: Imp. de D. José del Collado, 1821). Laqueur considers this work by Roussel to be one of the most representative examples of sexual dimorphism and of the biological interpretation present from the Enlightenment onwards. See Laqueur, *Making Sex*, p. 6. Consider the words of Roussel: 'Parece pues que el temperamento que se llama sanguíneo es en general el de las mugeres ... Unas fibras débiles y fáciles de moverse deben necesitar un género de sensibilidad viva pero pasagera ... Los sentimientos más disparatados se suceden en ellas con una rapidez que espanta, de suerte que no es raro verlas reír y llorar muchas veces en un mismo momento' (It would appear that the temperament that goes by the name of sanguineous is in general that of women ... Weak and easily mobile fibres must require a kind of sen-

sibility that is vivid but passing ... The most extravagant feelings occur in them such as a rapidity of response that shocks, so that it is not uncommon to see them laugh and cry at the same moment) (Roussel, *Sistema Físico y Moral de la Muger*, p. 54).

135. J. Capuron, *Tratado de las Enfermedades de las Mugeres desde la edad de la pubertad hasta la crítica inclusive*, 2 vols (Madrid: Imprenta Calle de la Greda, 1821).

136. 'A pesar de los escritos llenos de ideas juiciosas de los sabios Geoffroy de Saint-Hilaire y de otros, no es fácil esplicar la causa de otras muchas monstruosidades' (In spite of the many writings full of intelligent ideas of the wise Geoffroy de Saint Hilaire and others, it is not easy to explain the causes of other such monstrosities); M. Hurtado de Mendoza, *Instituciones de Medicina*, vol. I (Madrid: Sánchez, 1839), p. 125. Geoffroy Saint-Hilaire, together with Meckel, was the essential reference point for nineteenth-century Spanish anatomists in questions of embryology and teratology. Saint-Hilaire divided hermaphrodites into two large groups: those without excess in their sexual parts and those with. Among the first class, there were masculine, feminine, neutral and mixed. The neutral class offered a combination of the organs of both sexes such 'que la détermination du véritable sexe soit difficile ou même entièrement impossible'; I. Geoffroy Saint-Hilaire, *Histoire Générale et Particulière des anomalies de l'organization chez l'homme et les animaux. Traité de Tératologie*, vol. 2 (Paris: J. B. Baillière, 1836), p. 36. Neutral hermaphrodites lacked sexual differentiation to the degree that they were considered of no sex. The mixed hermaphrodite had the characteristics of both sexes in such a way that one part corresponded to one sex and the other part to the other sex and was thus a contrast to the 'disordered' neutral hermaphrodite. The second class was made up by hermaphrodites who did possess sexual parts in excess. These, in turn, were divided into masculine, feminine and bisexual. The latter possessed the sexual parts of both sexes. They could be imperfect if one set of genitalia or both were incomplete, or perfect if the genitalia were complete. Saint-Hilaire denied the existence of this sub-type: 'c'est à dire, la réunion d'un appareil mâle et d'un appareil femelle entièrement complets. Mais nous verrons que, malgré les nombreux témoignages consignés dans les ouvrages des anciens auteurs, l'observation et la théorie s'accordent pour démentir l'existence de ce dernier groupe' (that is to say, the coming together of an entirely complete male and female apparatus. Instead we see that, despite the numerous witnesses present in works of ancient authors, observation and theory are in agreement in denying the existence of this group) (Saint-Hilaire, *Histoire Générale*, p. 38). On the asymptotic condition of the 'perfect hermaphrodite' in Saint-Hilaire, see M. Tort, 'Le mixte et l'Occident. L'hermaphrodite entre le mythe et la science. Platon, Ovide, Isidore Geoffroy Saint Hilaire', in *La Raison Classificatoire* (Paris: Aubier Montaigne, 1989), pp. 175–203, on p. 197. On the significance of Saint-Hilaire for the Valencian anatomist Lorenzo Boscasa e Igual (1786–1857), see J. Arechaga Martínez, *La Anatomía Española en la Primera Mitad del Siglo XIX* (Granada: Universidad de Granada, 1977), p. 164. The reception of Meckel, whose classification of hermaphroditism drew in part on that of Saint-Hilaire (see Saint-Hilaire, *Histoire Générale*, p. 35), in Spain was greater than that of Saint-Hilaire, although the latter was mentioned favourably in Hurtado de Mendoza, Agapito Zuriaga y Clemente (1814–66) and Mariano López Mateos (1802–63), and not so favourably in Fabra y Soldevilla and Boscasa e Igual (see Arechaga Martínez, *La Anatomía Española*, p. 220).

137. 'Hermafroditismo o reunión de los dos sexos que comúnmente llaman hermafroditas, es una fábula transmitida de la antigüedad, en que en aquellos tiempos se carecía de los conocimientos anatómicos exactos, pues es imposible que en el hombre y en la numerosa familia de los animales de sangre roxa se verifique semejante unión. Las observaciones

exactas que se han podido recoger por los más distinguidos profesores no ofrecen testimonio alguno auténtico que lo confirme, y todos los hermafroditas que se han podido
ver hasta ahora, y de que hacen mención algunos autores, no han sido más que unos seres
mal conformados' (Hermaphroditism, or the uniting of the two sexes who are commonly
called hermaphrodites, is a fable transmitted from ancient times, times in which precise
anatomical knowledge was lacking, since it is impossible that in man and in the large
family of animals of red blood a similar kind of union occurs. Precise observations collected by distinguished professors have offered no authentic proof that would confirm
this, and all hermaphrodites that have been seen to date, and which have been mentioned
by certain authors, have been nothing but creatures poorly formed); A. Ballano, 'Hermafrodita', in *Diccionario de Medicina y Cirugía o Biblioteca Manual Médico-Quirúrgica*,
vol. 5 (Madrid: D. Francisco Martínez Dávila, 1817), pp. 102–3. This concept is prati-
cally identical in M. Hurtado de Mendoza, *Vocabulario Médico-Quirúrgico o Diccionario
de Medicina y Cirugía* (Madrid: Boix Editor, 1840), pp. 478–9. Hurtado de Mendoza
at the time still felt the need to reject the theory of hermaphroditism as the result of
a 'moral impression' during pregnancy. This argument 'aunque desgraciadamente sea el
más acreditado en el público, es el menos fundado de todos' (although it is unfortunately
the one most highly believed by the public, has the least foundation of all) (Hurtado
de Mendoza, *Instituciones de Medicina*, p. 125). On Hurtado de Mendoza's work, see
Arechaga Martínez, 'Manuel Hurtado de Mendoza (1780–85–1849)', in *La Anatomía
Española*, pp. 31–102. Hurtado was to state: 'La etimología de la palabra ... prueba que,
desde la más remota antigüedad, se ha creído en la existencia de estos seres quiméricos ...
La ignorancia y la credulidad aumentaron y perpetuaron este error de siglo en siglo, hasta
el punto que, en tiempos más modernos, se han visto personages graves, y aun médicos
que, engañados por apariencias, llevaron su absurdo hasta citar ejemplos de conversión
de muchachas en muchachos, a la época de la menstruación, o en la primera noche de
matrimonio' (The etymology of the word ... proves that, from the most ancient times,
the existence of these chimerical beings has been believed ... Ignorance and credulity
increased and perpetuated this error over centuries to the point that, in more modern
times, serious observers and indeed doctors who have been deceived by appearances have
cited absurd examples of conversion of girls into boys at the time of menstruation or on
the first night of matrimony); A. Hurtado de Mendoza, 'Hermafrodismo', in *Suplemento
al Diccionario de Medicina y Cirugía del Profesor D. Antonio Ballano*, vol. 3 (Madrid:
Imprenta de Brugada, 1823), p. 1135. De Viguera wrote that it was the scalpel that had
done the trick to disabuse people of this phenomenon: 'la brújula del escalpelo desen-
trañó por fin el simulacro del prodigio e hizo desaparecer lo maravilloso' (a firm hold
on the scalpel dispatched once and for all the simulacrum of the prodigy and made the
marvellous disappear) (de Viguera, 'Apuntes sobre el hermafrodismo', in *La Fisiología y
Patología de la Muger*, p. 116). See also the 'Apuntes sobre la metamorfosis sexual' by the
same author in *La Fisiología y Patología de la Muger*, pp. 126–9.

138. 'Debería borrarse del lenguage médico la palabra "hermafrodismo" siempre que se tratase
de la especie humana. Consecuente yo con esta opinión, no la usaré de manera alguna'
(The word 'hermaphrodism' should be expunged from the medical lexicon when referring to the human species. Consequent with this opinion, I shall by no means employ
the term); M. Orfila, *Tratado de Medicina Legal*, 4 vols (Madrid: Imp. de D. José María
Alonso, 1847), vol. 1, p. 188.

139. 'Por hermafroditismo en el hombre o la mujer se entiende aquella disposicion viciosa
de las partes genitales en la que el individuo parece ser de un sexo, a que realmente no

pertenece, o no se puede determinar cuál sea el verdadero sexo' (By hermaphroditism in man or woman we understand that vicious disposition of the genital parts that makes the individual appear to be of the sex to which he does not belong or for whom one cannot determine their real sex) (Mata, *Vademecum*, pp. 45–6).

140. Peiró and Rodrigo, *Elementos de Medicina y Cirugía Legal*, 3rd edn 1841, p. 9, wrote of the 'ignorance' and 'credulity' that engendered such thinking and action and condemned the 'letting of innocent blood' as a result. The Aragonese Pedro Miguel de Peiró was a doctor of law and became an emeritus member of the Academia Matritense de Jurisprudencia y Legislación. José Rodrigo, also Aragonese, was a doctor in medicine and surgery. This text, published in Zaragoza in 1832, was the first of its kind and became the manual that was used in all surgeons' colleges throughout Spain. For more details, see Pérez de Petinto and Bertomeu, 'Comienzo y actualidad (en 1951) de la trayectoria corporativa médico-forense', p. 6. The words of Peiró-Rodrigo mirror almost exactly – except in the two cases he mentions of the Scottish servant and the French woman – those written by Hurtado de Mendoza, 'Hermafrodismo', p. 1135. For the same kind of reasoning, see de Viguera, Apuntes sobre el hermafrodismo, in *La Fisiología y Patología de la Muger*, p. 116.

141. 'Los progresos de la anatomía y fisiología, señaladamente desde que se hace una aplicación exacta y rigurosa de las ciencias a la medicina legal, han hecho que se estudien con un cuidado particular los diferentes casos que se confundían en otro tiempo con la designación vaga de hermafrodismo, y que se fijen de un modo incontestable sobre esta materia los profesores del arte de curar y los jurisconsultos' (The progress of anatomy and physiology, particularly since the exact and rigorous application of these sciences within legal medicine, has meant that those cases that were confused in earlier times with the vague description of hermaphrodism have been studied with especial care) (Hurtado de Mendoza, 'Hermafrodismo', p. 1135).

142. S. N., 'Nueva aplicación del microscopio a los experimentos médico-legales', *Boletín de Medicina, Cirugía y Farmacia*, 2:66 (1841), p. 237. Cf. R. Cleminson and R. Medina Doménech, '¿Mujer u hombre? Hermafroditismo, tecnologías médicas e identificación del sexo en España, 1860–1925', *Dynamis*, 24 (2004), pp. 53–91, esp. pp. 80–4.

143. Dr. Henri Marc (1771–1840) was a decisive authority on the question. He established a series of rules for the diagnosis of true sex, and these were retained for decades. Marc believed that the possibility of reproduction among hermaphrodite individuals was something that divided medical opinion. See C. C. H. Marc, 'Hermaphrodite', in *Dictionnaire des Sciences Médicales par une Société de Médecins et de Chirurgiens, Vol. XXI: HEM–HUM* (Paris: C. L. F. Panckoucke, 1817), pp. 86–121. Dr Juan Mosácula (1794–1831), professor of physiology at the Colegio de San Carlos, denied the existence of hermaphroditism in humans; see J. Mosácula, *Elementos de Fisiología Especial o Humana*, vol. 2 (Madrid: Hijos de C. Piñuela, 1830), pp. 370–1. Other authors refer to the impossibility of reproduction in all cases: Ballano, 'Hermafrodita', p. 103; Hurtado de Mendoza, 'Hermafrodismo', p. 1136; Peiró and Rodrigo, *Elementos de Medicina y Cirugía Legal*, p. 9; Hurtado de Mendoza, *Vocabulario Médico-Quirúrgico*, p. 479; Orfila, *Tratado de Medicina Legal*, vol. 1, p. 188. Pedro Mata argued that even the 'neutral hermaphrodite' could be declared potent. No *a priori* statements were justified (Mata, *Vademecum*, p. 21).

144. P. F. Monlau, *Higiene del Matrimonio o el Libro de los casados* (1853; Madrid: M. Rivadeneyra, 1868), p. 158. Pedro Felipe Monlau (1808–71) was a member of the Consejo de Sanidad del Reino and the head of Spanish hygiene in the mid-nineteenth century. His *Higiene del Matrimonio* underwent numerous editions, and it became

extremely well known (republished seven times up to 1898 and translated into French in 1879). On Monlau, see M. Granjel, *Pedro Felipe Monlau y la Higiene Española del siglo XIX* (Salamanca: Universidad de Salamanca, 1983).

145. 'Diferentes hechos atestiguan que hay seres monstruosos que reunen los atributos de ambos sexos; y otros en quienes no se observa carácter ninguno distintivo: y esto es lo que ha hecho decir a Blumenback, a Meckel, a Geoffroy Saint Hilaire, que los dos sexos presentan en su estado primitivo, una sola y misma forma, y que solos los progresos del incremento son los que desenvuelven los caracteres propios de cada uno de ellos' (Different facts prove that there are monstrous beings that unite the attributes of both sexes, and others in whom no distinctive characteristic can be observed. This is what has driven Blumenback, Meckel and Geoffroy Saint-Hilaire to say the sexes in their primitive state represent a unique and single form and that it is only the progress of change that develops the characteristics of one or the other); M. Dany, 'Observación que puede servir para la historia del hermafrodismo', *Gaceta Médica de Madrid*, 1 (1835), p. 151. A similar take is seen in Hurtado de Mendoza, *Vocabulario Médico-Quirúrgico*, p. 478; Orfila, *Tratado de Medicina Legal*, vol. 1, p. 188; and Mata, *Vademecum*, p. 15. The lack of differentiation of infancy and old age is emphasized by Virey in his *Tratado Histórico y Fisiológico Completo*, p. 75, and by de Viguera in his 'Apuntes sobre el hermafrodismo', in *La Fisiología y Patología de la Muger*, p. 127.

146. G. Canguilhem, *The Normal and the Pathological*, trans. C. R. Fawcett and R. S. Cohen (New York: Zone Books, 1989), pp. 131–7.

147. The shift in medico-legal medicine on the question of hermaphroditism in France and its consequences for legislation are analysed by the University of Granada doctor José de Lletor Castroverde; see J. de Lletor Castroverde, *Repertorio Médico Extranjero*, vol. 5 (Madrid: Imprenta Real, 1835), p. 73. On the differences between vices of conformation and monstrosity in cases of hermaphroditism, see Orfila, *Tratado de Medicina Legal*, vol. 1, p. 193.

148. On Marc's criteria and their use in Spain, see Cleminson and Medina Doménech, '¿Mujer u hombre?'

149. Dreger, *Hermaphrodites*, pp. 139–66. On Klebs in Spain, see Cleminson and Medina Doménech, '¿Mujer u hombre?', pp. 79–80.

4 Hermaphroditism in Portugal

1. As S. Haliczer, 'The First Holocaust: The Inquisition and the Converted Jews of Spain and Portugal', in *Inquisition and Society in Early Modern Europe* (London and Sydney: Croom Helm, 1987), pp. 7–18, on p. 16 n. 6, has pointed out, many historians use the term 'marrano' despite its offensive origins. On the Arabic origins of the term *máhran*, that which is illicit, and its association with pigs, see I. S. Révah, 'Les Marranes', *Revue des Études Juives*, 118 (1959–60), pp. 29–77, on p. 30.

2. See also Saint Paula de Ávila, Eugenia and others mentioned in Chapter 1.

3. There is surprisingly little written about Wilgefortis. One major source is Friesen, *The Female Crucifix*; see also K. Coyne Kelly, *Performing Virginity and Testing Chastity in the Middle Ages* (London and New York: Routledge, 2000), pp. 52–3 and 93.

4. A. Lorenzo Vélez, 'El motivo de "La Mujer Disfrazada de Varón" en la tradición oral moderna', *Revista de Folklore*, 17a:194 (1997), pp. 39–59, citing V. Vega, *Diccionario Ilustrado de Rarezas, Inverosimilitudes y Curiosidades*, 4th edn (Barcelona: Gustavo Gili, 1971), p. 538.

5. D. Nunez do Leão, *Descripção do Reino de Portugal* (Lisbon: Iorge Rodriguez, 1610), Chapter 79, 'Do valor & animo de molheres Portuguesas', fols 144v–151v, on fols 148v–150v. The later spelling of Antónia, with a written accent, is the one adopted here. The author's name has also been spelled Nunes de Leão. See Boxer, *Mary and Misogyny*, pp. 12–16, for the broader involvement of Portuguese women in North Africa.

6. See Frei João de São Pedro (Damião de Froes Perym), *Theatro Heroino, Vol. I: Abecedario historico, e catalogo das mulheres ilustres em armas, letras, acçoens heroicas, e artes liberaes* (Lisbon: Theotonio Antunes Lima, 1736), pp. 54–7; A. da Costa, *A mulher em Portugal* (Lisbon: Typ. da Companhia Nacional Editora, 1892), pp. 33–5; Conde de Sabugosa, *Neves de Antanho*, 3rd edn (Lisbon: Livraria Bertrand, 1919), chapter entitled 'Antónia Rodrigues. Amazona de Mazagão', pp. 257–80; and Pires de Lima, 'Hermafroditismo e Inter-Sexualidade', pp. 472–3.

7. Nunez do Leão, *Descripção do Reino de Portugal*, fol. 144r. In accordance with the original account, the pronoun used here to introduce Antónia/o shifts from female to male and back to female again.

8. This is asserted by Nunez do Leão in ibid., de Froes Perym, *Theatro Heroino*, p. 56, and Sabugosa, *Neves de Antanho*, p. 280. The document recording this development is in the Arquivo da Torre do Tombo, Chancel. de Felip, II, Doaç, Liv, 12fl, 18v.

9. See, for example, Burshatin, 'Written on the Body', Vázquez García and Cleminson, 'Subjectivities in Transition', and other cases discussed in this volume.

10. De Froes Perym, *Theatro Heroino*, p. 54, refers to her buying 'hum vestido dos que uzavaõ no mar os homens da sua terra' (a vestment that the men of her land used on the seas), then 'cortou os cabellos, e vestio os habitos de varão' (she cut her hair and dressed in the habit of a man).

11. We can contrast the fate of Antónia with those interrogated for sodomy by the Inquisition in North Africa some years before Antónia's adventures. Rafael de Oliva, a New Christian and resident of Mazagão, was tried for sodomy in 1587, and his trial can be seen in Arquivo Nacional da Torre do Tombo, Lisbon, Portugal (hereafter ANTT), Inquisição de Lisboa, processo n° 1679; the trial of the 'Moor slave', António de Brito from Tetuan, in the late 1540s can be seen in ANTT, Inquisição de Lisboa, processo n° 4807.

12. Immortalized in T. Artus, *L'Isle des Hermaphrodites, nouvellement descouverte* (c. 1605). See Long, *Hermaphrodites in Renaissance Europe*. Cf. I. Teixeira Rodrigues, 'Amato Lusitano e as perturbações sexuais. Algumas contribuições para uma nova perspectiva de análise das *Centúrias de Curas Medicinais*' (unpublished doctoral thesis, Universidade de Trás-os-Montes e Alto Douro, 2005), p. 256.

13. See Visconde de Almeida-Garrett, *Romanceiro. III. Romances cavalherescos antigos* (Lisbon: Viuva Bertrand e Filhos, 1863), ch. 24, 'A Donzella que vai á guerra', pp. 69–82. The Castilian origins of this romance have been disputed. For more details, see Lorenzo Vélez, 'El motivo de "La Mujer Disfrazada de Varón"'. A brief account of the 'girl who went to war' instead of her father is provided by G. Margouliès, 'Deux poèmes sur la jeune fille partie à la guerre por remplacer son père', *Revue de Littérature Comparée*, 8 (1928), pp. 304–9, on pp. 304–7, together with its French version, cited in W. Schleiner, 'Le feu caché: Homosocial Bonds Between Women in a Renaissance Romance', *Renaissance Quarterly*, 45:2 (1992), pp. 293–311, on p. 294 n. 3.

14. Pires de Lima, 'Hermafroditismo e Inter-Sexualidade', p. 473.

15. A further development was provided by Joaquim Alberto Pires de Lima's son, Fernando de Castro Pires de Lima, in his *A mulher vestida de homem. Contribuição para o estudo*

do romance 'A donzela que vai á guerra' (Coimbra: Fundação Nacional para a Alegria no Trabalho, 1958).

16. Soyer, *Ambiguous Gender*, p. 20.
17. E. A. Lehfeldt, 'Ideal Men: Masculinity and Decline in Seventeenth-Century Spain', *Renaissance Quarterly*, 61:2 (2008), pp. 463–94, on p. 464.
18. Ibid., p. 466.
19. J. de Santa María, *República y policía Christiana* (Lisbon, 1621), cited in Soyer, *Ambiguous Gender*, p. 19 n. 6.
20. Lehfeldt, 'Ideal Men', pp. 481–5.
21. L. Mott, 'Pagode português: a subcultura *gay* em Portugal nos tempos inquisitoriais', *Revista Ciência e Cultura*, 40:2 (1988), pp. 120–39, on p. 121.
22. Soyer, *Ambiguous Gender*, p. 20. See also A. A. de Aguiar, 'Crimes e delitos sexuais em Portugal na época das Ordenações (sexualidade anormal)', *Archivo de Medicina Legal*, 1:2 (1930), pp. 118–44, where the author refers to the legislation on what he terms 'transvestitism' on pp. 133–4.
23. The case forms part of the Inquisition archive at the ANTT, Inquisição de Lisboa, processo nº 10,868, discussed in Soyer, *Ambiguous Gender*, pp. 21–2.
24. Soyer, *Ambiguous Gender*, p. 21.
25. Ibid., p. 22. The case is preserved as ANTT, Inquisição de Lisboa, processo nº 5,947. Mott, 'Pagode português', p. 129, notes that the first case of 'transvestism' to come to the attention of the Inquisition in Portugal was that of the Benin-born slave, Antonio, in 1556. On this case, see also J. H. Sweet, 'Male Homosexuality and Spiritism in the African Diaspora: The Legacies of a Link', *Journal of the History of Sexuality*, 7:2 (1996), pp. 184–202, on p. 194 n. 33. Antonio was exiled to Brazil, where he worked as a female prostitute.
26. Y. H. Yerushalmi, *From Spanish Court to Italian Ghetto. Isaac Cardoso: A Study in Seventeenth-Century Marranism and Jewish Apologetics* (New York and London: Columbia University Press, 1971), pp. 2–6. On the origins of the Inquisition in Portugal, see A. Herculano, *Historia da origem e estabelecimento da Inquisição em Portugal*, ed. D. Lopes, 3 vols (Rio de Janeiro, São Paulo, Belo Horizonte and Lisbon: Editoria Paulo de Azevedo/ Livraria Bertrand, 1925), vol. 1, pp. 107–99, on the Jews in Portugal.
27. Kamen, *The Disinherited*, p. 10.
28. Ibid., pp. 13–15.
29. Of the 23,068 individuals penanced between 1536 and 1732, nine-tenths were New Christian Judaizers, according to Haliczer, 'The First Holocaust', p. 16, citing J. Lucio d'Azevedo, *Historia dos Christãos Novos Portugueses* (Lisbon: Livraria Clássica, 1922), p. 237 [in reality, p. 489].
30. Kamen, *The Disinherited*, p. 28.
31. Ibid., pp. 28–30.
32. See E. Shohat, *Dangerous Liaisons: Gender, Nations, and Postcolonial Perspectives* (Minneapolis, MN and London: University of Minnesota Press, 1997), p. 92; and P. Seed, *Ceremonies of Possession in Europe's Conquest of the New World, 1492–1640* (Cambridge: Cambridge University Press, 1995).
33. Teixeira Rodrigues, 'Amato Lusitano', pp. 69–71.
34. L. Thorndike, *A History of Magic and Experimental Science, Vols. 7 & 8: The Seventeenth Century* (New York and London: Colombia University Press, 1958).
35. M. Menéndez Pelayo, *Historia de los heterodoxos españoles*, 3 vols (Madrid: Librería Católica de San José, 1880), vol. 2, pp. 586–619, discusses the important role of Portuguese-origin Jews in science.

36. Efron, 'Nature, Human Nature, and Jewish Nature'.
37. H. Friedenwald, *Jewish Luminaries in Medical History* (Baltimore, MD: Johns Hopkins University Press, 1946), p. 15.
38. Paré, *On Monsters and Marvels*, p. 31, where Lusitanus's comments on the case of Maria Pacheca are recorded (discussed below).
39. Friedenwald, *Jewish Luminaries*, pp. 16–17.
40. Teixeira Rodrigues, 'Amato Lusitano', p. 191, drawing on the work of C. Pinto-Correia and J. P. Sousa Dias, *Assim na Terra como no Céu. Ciência, Religião e Estruturação do Pensamento Ocidental* (Lisbon: Relógio D'Água, 2003), p. 359, refers to the censorship of Amato's work in 1581, 1612, 1632, 1640 and 1707. See also J. Pardo Tomás, *Ciencia y censura. La Inquisición Española y los libros científicos en los siglos XVI y XVII* (CSIC: Madrid, 1992); and specifically on Lusitanus and the expurgation of his discussion of false conception by a nun in the *Centuriae*, see D. Front, 'The Expurgation of Medical Books in Sixteenth-Century Spain', *Bulletin of the History of Medicine*, 75:2 (2001), pp. 290–6. On the question of the thought of Albucasis with respect to hermaphroditism, see Cleminson and Vázquez García, *Hermaphroditism*, p. 23 n. 30. On the 'rediscovery' of Greek philosophical thought in general and on the body in particular via the Islamic presence in Spain and Italy and its translation into Latin via Arabic, see D. Jacquart, 'Influence de la médicine arabe en Occident médiéval', in R. Rashed (ed), *Histoire des sciences arabes, vol. III: Technologie, alchimie et sciences de la vie* (Paris: Éditions du Seuil, 1997), pp. 213–32; R. Fletcher, *Moorish Spain* (London: Phoenix, 1998), pp. 8–9 and 133; and K. Crawford, 'The Good, the Bad, and the Textual: Approaches to the Study of the Body and Sexuality, 1500–1750', in S. Toulalan and K. Fisher (eds), *The Routledge History of Sex and the Body 1500 to the Present* (London and New York: Routledge, 2013), pp. 23–37, on p. 26.
41. A. Lusitani, *Curationum medicinalium. Centuriae II*, 'Curatio XXXIX' (Lugduni [Lyon]: Gulielmum Rouillium, 1567), pp. 553–4.
42. Teixeira Rodrigues, 'Amato Lusitano', p. 193.
43. The figure of 17 per cent comes from ibid., p. 196.
44. Ibid., p. 199, on discussing IV *Centúria*, Cure XIII.
45. Ibid., p. 213, in her discussion of the author's I *Centúria*, Cure LXX.
46. See Paré, *On Monsters and Marvels*, pp. 26–33, where sex change and hermaphroditism were readily accepted.
47. Teixeira Rodrigues, 'Amato Lusitano', p. 249. See Lusitani, *Curationum medicinalium. Centuriae I*, Curatio XXIII, p. 169.
48. Teixeira Rodrigues, 'Amato Lusitano', p. 250.
49. Ibid., p. 251, citing V *Centúria*, Cura III.
50. See, for example, Peramato, *Opera Medicinalia*, second treatise, fol. 117r, and I. Cardoso, *Philosophia libera in septem libros distributa* (Venice: Bertanorum sumptibus, 1673), pp. 461–2, at http://books.google.co.uk/books?id=c0BfJaN5mA4C&printsec=frontcov er&hl=es&source=gbs_ge_summary_r&cad=0#v=onepage&q&f=false [accessed 22 August 2013].
51. Pires de Lima, 'Hermafroditismo e Inter-Sexualidade', p. 474; M. Lemos, *Amato Lusitano: A sua vida e a sua obra* (Oporto: Typ. da Encyclopedia Portugueza Ilustrada, 1907), p. 44.
52. Lusitani, *Curationum medicinalium*, pp. 553–4.
53. We have consulted the third edition: Castro Lusitani, 'De hermaphroditis', in *De universa muliebrium morborum medicina*, pp. 144–7.

54. Friedenwald, *Jewish Luminaries*, p. 55, gives his date of death as 1627; other sources give later dates. W. Schleiner, 'Cross-Dressing, Gender Errors, and Sexual Taboos in Renaissance Literature', in S. P. Ramet (ed.), *Gender Reversals and Gender Cultures: Anthropological and Historical Perspectives* (London and New York: Routledge, 1996), pp. 92–104, on p. 102 n. 6, notes that he argued sex mutation was possible. We have consulted *Tractatus de natura muliebri* (Hamburg: Sande, 1608), pp. 55–63, where 'sex permutations' are discussed. See http://search.ugent.be/meercat/x/bkt01?q=900000147674 [accessed 22 August 2013].

55. For a discussion of the history of Portuguese monsters and the science of teratology, see the work of J. A. Pires de Lima, 'Contribuïção para a história da Teratologia portuguesa. Monstros duplos autositários', *Arquivo de Anatomia e Antropologia*, 18 (1937), pp. 53–60.

56. Castro Lusitani, *De universa muliebrium*, p. 144.

57. Peramato, *Opera Medicinalia*, second treatise, section on 'Pueri et Puerperae regimine', and specifically Chapter VIII, 'De puero qui nascitur mutilus', fol. 117r–v. On Peramato, see A. Hernández Morejón, *Historia bibliográfica de la medicina española*, vol. 5 (Madrid: Imprenta de la Viuda de Jordán é Hijos, 1847), pp. 20–7, who notes that he studied at Alcalá and Salamanca; and Chinchilla, *Anales históricos de la medicina en general*, vol. 2, pp. 77–81.

58. Castro Lusitani, *De universa muliebrium*, pp. 144–5.

59. Ibid., p. 145.

60. Ibid., p. 146.

61. E. Rodrigo de Castro, *Tractatus de natura muliebri* (Hamburg: Sande, 1608), p. 55.

62. Ibid., p. 56.

63. Ibid., p. 57.

64. Z. Lusitani, *Operum Tomos Primus in quo de Medicorum Principum Historia* (Lyon: Joanni Antonii Huguetan & Marci Antonii Ravaud, 1649), p. 467. See also Friedenwald, *Jewish Luminaries*, p. 154.

65. Zacuto also referred to the case of Maria Pacheca as detailed by Amatus Lusitanus to back up his ideas.

66. M. Donato, *De medica historia mirabili* (Venice, 1588), as part of the discussion on monsters and hermaphrodites; on p. 298 the case of Maria Pacheca is also mentioned.

67. Yerushalmi, *From Spanish Court to Italian Ghetto*, pp. 249–50.

68. Cardoso, *Philosophia libera*, pp. 461–4.

69. Ibid., pp. 464–6.

70. Ibid., p. 461.

71. Ibid., pp. 461–2.

72. Ibid., p. 462.

73. Ibid.

74. Ibid., p. 463. Del Río and Riolan were employed to back this negation.

75. Ibid., p. 464.

76. Ibid., p. 465.

77. Friedenwald, *Jewish Luminaries*, p. 55, for example, notes that Zacutus spoke of Stephan Rodrigo de Castro as the 'Phoenix of medicine'.

78. Pires de Lima, 'Hermafroditismo e Inter-Sexualidade', p. 474.

79. Ibid. See Padre A. Vieira, *Arte de Furtar* (Lisbon: Livraria Peninsular Editora, 1937), p. 134, where the original reads 'Nem é deformidade nem impossível que a fêmea represente sexo de varão ... porque a natureza faz muitas vêzes das fêmeas machos'.

80. G. dos Reis Franco, *Elysius iucundarum quaestionum campus* (Brussels: Typis & sumpti-bus Francisci Vivien, 1661).

81. Pires de Lima, 'Hermafroditismo e Inter-Sexualidade', p. 474. No actual work by this author is referred to.

82. Reis Franco, *Elysius iucundarum quaestionum campus*, pp. 349–58.

83. Ibid., p. 350.

84. Ibid., p. 351.

85. Ibid., p. 352.

86. Ibid., p. 353. It was only later, on p. 355 (printed as p. 353 in the copy consulted), that he detailed the four categories of hermaphrodite, *contra* those who argued for three: the first type, the male hermaphrodite, was a practically perfect male, but possessed an indenta-tion in the perineum that had the form of a vulva; the second, the female hermaphrodite, despite possessing a vulva 'in accordance with nature', also had a 'fleshy body' that appeared to be a penis but which was imperforate, and no testicles were found; the third possessed the appearance of the two sexes in contrary places and could not emit semen or be impregnated; the fourth possessed the genitalia of both sexes and was potent.

87. Ibid., p. 353.

88. Ibid., p. 354.

89. Ibid., p. 353 [p. 355].

90. Ibid., p. 356.

91. Ibid., pp. 356–7.

92. Ibid., p. 358.

93. Ibid. For an analysis of such a case in France, that of Marguerite Malaure in the late sev-enteenth century, see Graille, *Le Troisième sexe*, pp. 119–23.

94. For a detailed and reflexive account of the available sources and the methodology to be employed with respect to Inquisition sources, although mainly confined to Spain, consult Soyer, 'Inquisitors and Hermaphrodites', in *Ambiguous Gender*, pp. 11–16 and 50–95. For an analysis more relevant to Portugal and Brazil, see D. Higgs, 'Tales of Two Carmelites: Inquisitorial Narratives from Portugal and Brazil', in P. Sigal (ed.), *Infamous Desire: Male Homosexuality in Colonial Latin America* (Chicago, IL: Chicago Univer-sity Press, 2003), pp. 152–67.

95. Soyer, *Ambiguous Gender*, pp. 20–1.

96. Discussed briefly in Sweet, 'Male Homosexuality', pp. 194–6, and mentioned in Soyer, *Ambiguous Gender*, pp. 10–11. See ANTT, Inquisição de Évora, processo nº 4,745.

97. Soyer, *Ambiguous Gender*, pp. 125–80.

98. Ibid., pp. 181–209.

99. Specifically on this association, see J. Richards, *Sex, Dissidence, and Damnation: Minor-ity Groups in the Middle Ages* (London and New York: Routledge, 1991), pp. 20–1, 59–62 and 84.

100. Mott, 'Pagode português', p. 122; L. Mott and A. Assunção, 'Love's Labors Lost: Five Letters from a Seventeenth-Century Portuguese Sodomite', in K. Gerard and G. Hekma (eds), *The Pursuit of Sodomy: Male Homosexuality in Renaissance and Enlightenment Europe* (New York and London: Harrington Park Press, 1989), pp. 91–101.

101. De Aguiar, 'Pseudo-hermafroditismo feminino'.

102. The Santa Cruz e Luz monastery was established in 1525 and was first populated by nuns from the Santa Mónica monstery in Évora; while awaiting the construction of the new Vila Viçosa establishment, they occupied the Chagas convent in the same town until 1530. The monastery was closed in 1883. See J. L. Inglês Fontes, J. Bastos Serra and M.

F. Andrade, *Inventário dos Fundos Monástico-Conventuais da Biblioteca Pública de Évora* (Lisbon and Évora: Edições Colibri/Universidade de Évora, 2010), p. 252. We would like to thank Dr María Zozaya (University of Évora) for her assistance in tracing materials pertinent to this convent.

103. De Aguiar, 'Pseudo-hermafroditismo feminino', p. 433.
104. Ibid., p. 434.
105. Ibid.
106. Ibid., p. 435.
107. Ibid.
108. Ibid.
109. Ibid.
110. Ibid., p. 436. Cf. the discussion on similar cases of papal dispensation in Spain in Chapter 2.
111. Soyer, *Ambiguous Gender*, pp. 20–1. See ANTT, Inquisição de Coimbra, processo nº 7,083.
112. Soyer, *Ambiguous Gender*, p. 21.
113. Sweet, 'Male Homosexuality', p. 194.
114. ANTT, Inquisição de Évora, processo nº 4,745, fol. 274v.
115. Ibid., fol. 279v and fol. 283v.
116. Ibid., fol. 279r and fol. 285r; note in margin that Estêvão Luís was famed for being an 'embosteiro, feitisseiro, bruxo e macho e femea' (cheat, conjurer, witch and man and woman).
117. Ibid., fol. 280v.
118. Ibid., fol. 314v.
119. Ibid., fol. 269v and fol. 314r. Cf. Sweet, 'Male Homosexuality', p. 195.
120. Sweet, 'Male Homosexuality', p. 195.
121. Soyer, *Ambiguous Gender*, pp. 125–80. All citations in the Portuguese for this case are taken from Soyer's account. Our translation.
122. Ibid., p. 127.
123. Ibid., p. 132 n. 18.
124. Ibid., p. 135 n. 25.
125. Ibid., pp. 140–2.
126. Ibid., p. 162 n. 81.
127. Ibid., p. 176 n. 111.
128. Ibid., p. 177.
129. Ibid., p. 178.
130. On the lack of cited sources, see ibid., p. 178.
131. The case is amply covered in ibid., pp. 181–209.
132. Ibid., p. 184.
133. Ibid., p. 307.
134. Ibid., p. 188.
135. Ibid.
136. Ibid., p. 191 n. 16.
137. Both quotations from ibid., p. 192 n. 19.
138. Ibid., p. 203 n. 41.
139. Ibid., p. 195.
140. Ibid., p. 126.
141. This important point on the significance of cases of hermaphroditism in periods of religious conflict is made by Graille, *Le Troisième sexe*, p. 118.
142. Soyer, *Ambiguous Gender*, p. 170.

Conclusion

1. M. Foucault, *El orden del discurso*, trans. A. González Troyano (1970; Barcelona: Tusquets, 2002), p. 53.
2. Marchetti, *L'invenzione della bisessualità*.
3. B. de Sousa Santos, 'Between Prospero and Caliban: Colonialism, Post-Colonialism, and Inter-identity', *Luso-Brazilian Review*, 39:2 (2002), pp. 9–43.
4. Teixeira Rodrigues, 'Amato Lusitano'.
5. Cleminson and Vázquez García, *Hermaphroditism*, p. 23 n. 30; Jacquart, 'Influence de la médicine arabe'; Fletcher, *Moorish Spain*, pp. 8–9 and p. 133.
6. Cadden, *Meanings of Sex Difference*, p. 3.
7. Daston and Park, 'The Hermaphrodite and the Orders of Nature' (1996), pp. 117–36, on p. 118.
8. Soyer, *Ambiguous Gender*.
9. Foucault, *El orden del discurso*, p. 36.
10. Pires de Lima, 'Contribuïção para a história da Teratologia portuguesa'; Pires de Lima, 'Hermafroditismo e Inter-Sexualidade'.
11. See, for example, the enduring interest in cases of 'hermaphroditism' in Portugal in 1927: 'Mulher-homem chegada a Lisboa, e que a Polícia levou ao Governo Civil (tendo declarado o médico tratar-se de um caso de hermafroditismo)', ANTT reference PT/TT/EPJS/SF/001-001/0195/0331B, at http://digitarq.dgarq.gov.pt/details?id=1206869 [accessed 21 August 2013]. A hermaphrodite who possessed 'les organes des deux sexes dans le plus haut degré de perfection que l'on ait vu' (the organs of both sexes in what was the highest degree of perfection even seen) was identified in Lisbon in 1807, according to the medical doctor Béclard, cited in Marc, 'Hermaphrodite', p. 110.
12. José Ortega y Gasset, cited in M. Zambrano, *Persona y democracia. La historia sacrificial* (1958; Madrid: Ediciones Siruela, 1996), p. 84.

WORKS CITED

Agamben, G., *Homo Sacer: Sovereign Power and Bare Life*, trans. D. Heller-Roazen (Stanford, CA: Stanford University Press, 1998).

Aguiar, A. A. de, 'Pseudo-hermafroditismo feminino (caso português do século XVII)', *Archivo de Medicina Legal*, 2:4 supplement, 1923–5 (1928), pp. 432–6.

—, 'Crimes e delitos sexuais em Portugal na época das Ordenações (sexualidade anormal)', *Archivo de Medicina Legal*, 1:2 (1930), pp. 118–44.

Alberti López, L., *La Anatomía y los Anatomistas Españoles del Renacimiento* (Madrid: CSIC, 1948).

Alcalá Flecha, R., *Literatura e Ideología en el Arte de Goya* (Zaragoza: Diputación General de Aragón, 1988).

Albucasis, *On Surgery and Instruments*, ed. M. S. Spink and G. L. Lewis (London: Wellcome Institute of the History of Medicine, 1973).

Alcalá Galán, M., 'El andrógino de Francisco de Lugo y Ávila: discurso científico y ambigüedad erótica', *e-Humanista*, 15 (2010), pp. 107–35.

Alemán, M., *Guzmán de Alfarache*, in A. Valbuena y Prat (ed.), *La Novela Picaresca Española* (Madrid: Aguilar, 1968), pp. 159–637.

Alfonso X El Sabio, *Las Siete Partidas ... con las variantes de más interés y con la glosa del Licenciado Gregorio López*, ed. I. Samponts, vol. 4, Partida VI, Tit. 1, Ley 10 (Barcelona: Imprenta de Antonio Bergnés, 1844).

Almeida-Garrett, Visconde de, *Romanceiro. III. Romances cavalherescos antigos* (Lisbon: Viuva Bertrand e Filhos, 1863).

Alvarez de Miravall, B., *Libro intitulado La Conservación de la Salud del Cuerpo y del Alma* (Medina del Campo: S. del Canto, 1597).

Amar y Borbón, J., *Discurso sobre la Educación Física y Moral de las Mujeres* (1790; Madrid: Cátedra, 1994).

Ampudia de Haro, F., *Las Bridas de la Conducta. Una aproximación al proceso civilizatorio español* (Madrid: CIS, 2007).

Angelides, S., *A History of Bisexuality* (Chicago, IL: University of Chicago Press, 2001).

Arechaga Martínez, J., *La Anatomía Española en la Primera Mitad del Siglo XIX* (Granada: Universidad de Granada, 1977).

Arenas, A., 'La increíble vida de Elena/o de Céspedes', *Ideal Digital*, 14 August 2006, at http://www.ideal.es/granada/pg060814/prensa/noticias/Vivir/200608/14/ALM-SOC-074.html [accessed 22 August 2013].

Aresti Esteban, N., 'The Gendered Identities of the "Lieutenant Nun": Rethinking the Story of a Female Warrior in Early Modern Spain', *Gender and History*, 19:3 (2007), pp. 401–18.

Aristotle, *Generation of Animals*, Book IV, trans. A. L. Peck (London and Cambridge, MA: William Heinemann/Harvard University Press, 1943).

Arquivo Nacional da Torre do Tombo, Lisbon, Portugal: Inquisição de Coimbra; Inquisição de Évora; Inquisição de Lisboa.

Ashcomb, B. B., 'Concerning "la mujer en hábito de hombre" in the Comedia', *Hispanic Review*, 28:1 (1960), pp. 43–62.

Azevedo, J. Lucio d', *Historia dos Christãos Novos Portugueses* (Lisbon: Livraria Clássica, 1922).

Azpilcueta, M. de, *Manual de confessores et penitentes* (Coimbra: Ioan de Barnyra, 1549).

—, *Enchiridion sive manuale confessariorum ac poenitentiam* (Paris: Imp. Guillielmi Rovilli, 1587).

Bakhtin, M., *La Cultura Popular en la Edad Media y en el Renacimiento* (Madrid: Alianza Universidad, 1987).

Ballano, A., 'Hermafrodita', in *Diccionario de Medicina y Cirugía o Biblioteca Manual Médico-Quirúrgica*, vol. 5 (Madrid: D. Francisco Martínez Dávila, 1817), pp. 102–3.

Barbazza, M. C., 'Un caso de Subversión Social: el proceso de Elena de Céspedes (1587–1589)', *Criticón*, 26 (1984), pp. 17–40.

Barco y Gasca, A. J. del, 'Examen crítico de una rara transmutación de sexos en persona del femenino', in *Cartas Familiares, Varias y Curiosas, dispuestas para honesta diversión* (manuscript), vol. 3, letter 29 (Seville: Biblioteca de la Real Academia de Buenas Letras, 1770–1), fols 199–221.

Barret-Kriegel, B., *La défaite de l'érudition* (Paris: PUF, 1988).

Bento, B., 'La producción del cuerpo dimórfico: transexualidad e historia', *Anuario de Hojas de Warmi*, 15 (2010), pp. 1–19.

Bertelli, S., *Rebeldes, libertinos y ortodoxos en el Barroco* (Barcelona: Península, 1984).

Beusterien, J. L., 'Jewish Male Menstruation in Seventeenth-Century Spain', *Bulletin of the History of Medicine*, 73:3 (1999), pp. 447–56.

Bloom, A., and S. Benardete (eds), *Plato's 'Symposium'*, trans. S. Benardete (Chicago, IL and London: University of Chicago Press, 2001).

Boltanski, L., *Los Usos Sociales del Cuerpo* (Buenos Aires: Ediciones Periferia, 1975).

Bolufer Peruga, M., 'Lo Íntimo, lo Doméstico y lo Público. Representaciones y Estilos de Vida en la España Ilustrada', *Studia Histórica, Historia Moderna*, 19 (1998), pp. 85–116.

—, *Mujer e Ilustración. La construcción de la feminidad en la España del siglo XVIII* (Valencia: Institució Alfons el Magnànim, 1998).

—, 'Cos femení, cos social. Apunts d'historiografía sobre els sabers mèdics i la construcció cultural d'identitats sexuades (segles XVI–XIX)', *Afers. Full de Recerca i Pensament*, 33–4 (1999), pp. 531–50.

—, 'Galerías de "Mujeres Ilustres" o el sinuoso camino de la excepción a la norma cotidiana (ss. XV–XVIII)', *Hispania*, 60:204 (2000), pp. 181–224.

—, 'Literatura encarnada: modelos de corporalidad femenina en la Edad Moderna', in S. Mettalía and N. Girona (eds), *Aún y más allá: mujeres y discursos* (Caracas: Ex Cultura, 2002), pp. 205–15.

Borrachero Mendíbil, A., 'Catalina de Erauso ante el patriarcado colonial: un estudio de *Vida i sucesos de la Monja Alférez*', *Bulletin of Hispanic Studies*, 83:6 (2006), pp. 485–95.

Boswell, J., *Christianity, Social Tolerance and Homosexuality: Gay People in Western Europe from the Beginning of the Christian Era to the Fourteenth Century* (Chicago, IL and London: Chicago University Press, 1980).

Boucé, P. G., 'Imagination, Pregnant Women and Monsters in Eighteenth-Century England and France', in G. S. Rousseau and R. Porter (eds), *Sexual Underworlds of the Enlightenment* (Manchester: Manchester University Press, 1987), pp. 86–100.

Bourdieu, P., *La domination masculine* (Paris: Raisons d'Agir, 1998).

—, 'Les conditions sociales de la circulation internationale des idées', *Actes de la Recherche en Sciences Sociales*, 145 (2002), pp. 3–8.

—, *Sur l'État. Cours au Collège de France 1989–1992* (Paris: Raisons d'Agir, 2012).

Bouza, F., *Locos, enanos y hombres de placer en la Corte de los Austrias* (Madrid: Temas de Hoy, 1991).

Bovistuau, P., C. Tesserant and F. Belleforest, *Historias prodigiosas y maravillosas* (Madrid: Imp. Luis Sánchez, 1603).

Boxer, C. R., *Mary and Misogyny: Women in Iberian Expansion Overseas 1415–1815* (London: Duckworth, 1975).

Boylan, M., 'The Galenic and Hippocratic Challenges to Aristotle's Conception Theory', *Journal of the History of Biology*, 17:1 (1984), pp. 83–112.

Bradbury, G., 'Irregular Sexuality in the Spanish "Comedia"', *Modern Language Review*, 76:3 (1981), pp. 566–80.

Bravo de Sobremonte, G., 'Promptuarium XXIV', 'De sexus mutatione' and 'De hermaphroditis', in *Operum Medicinalium. Tomus Tertius* (Lyon: L. Arnaud, 1671), pp. 246–9 and pp. 249–50.

—, 'Resolutio I. Utrum sexus transmutatio permitatur naturae', in *Operum Medicinalium. Tomus Quartus: Tres Disputationes Complectens* (Lyon: L. Arnaud, 1679), p. 198.

Bravo-Villasante, C., *La Mujer Vestida de Hombre en el Teatro Español (siglos XVI–XVII)* (1955; Madrid: Mayo de Oro, 1988).

Brisson, L., *Sexual Ambivalence: Androgyny and Hermaphroditism in Graeco-Roman Antiquity* (Berkeley, CA, Los Angeles, CA and London: University of California Press, 2002).

Brown, J., *Sor Benedetta, entre sainte et lesbienne* (Paris: Gallimard, 1986).

Brown, P., *The Body and Society: Men, Women and Sexual Renunciation in Early Christianity* (New York: Columbia University Press, 1988).

Bueno, L., 'Identidad sexual del personaje en el teatro aurisecular', in A. Ceballos, R. Espejo and B. Muñoz (eds), *El teatro del género. El género del teatro. Las artes escénicas y la representación de la identidad sexual* (Madrid: Fundamentos, 2009), pp. 41–67.

Bullough, V. L., and B. Bullough, *Cross Dressing, Sex, and Gender* (Philadelphia, PA: University of Pennsylvania Press, 1993),

Burgos, E., 'Deconstrucción y subversión', in P. Soley-Beltran and L. Sabsay (eds), *Judith Butler en disputa. Lecturas sobre la performatividad* (Madrid and Barcelona: Egales, 2012), pp. 101–34.

Burke, P., *Varieties of Cultural History* (Ithaca, NY: Cornell University Press, 1997).

Burshatin, I., 'Elena alias Eleno. Genders, Sexualities and "Race" in the Mirror of Natural History in Sixteenth-Century Spain', in S. P. Ramet (ed.), *Gender Reversals and Gender Cultures: Anthropological and Historical Perspectives* (New York: Routledge, 1996), pp. 105–22.

—, 'Written on the Body: Slave or Hermaphrodite in Sixteenth-Century Spain', in J. Blackmore and G. S. Hutcheson (eds), *Queer Iberia: Sexualities, Cultures and Crossings from the Middle Ages to the Renaissance* (Durham, NC: Duke University Press, 1999), pp. 420–56.

Butler, J., *Gender Trouble: Feminism and the Subversion of Identity* (New York: Routledge, 1990).

—, *Bodies that Matter: On the Discursive Limits of 'Sex'* (New York and London: Routledge, 1993).

—, *Undoing Gender* (New York and London: Routledge, 2004).

Cadalso, J. de, *Cartas Marruecas* (1789; Madrid: Imprenta de Sancha, 1793).

Cadden, J., *Meanings of Sex Difference in the Middle Ages: Medicine, Science, and Culture* (Cambridge: Cambridge University Press, 1993).

Calabrese, O., 'Neobarroco', in F. Jarauta (ed.), *Otra Mirada sobre la Época* (Murcia: Colegio Oficial de Aparejadores y Arquitectos Técnicos/Cajamurcia, 1994), pp. 261–2.

Campillo, A., *La fuerza de la razón. Guerra, Estado y ciencia en el Renacimiento* (Murcia: Edit. um, 2008).

Campos y Fernández de Sevilla, F. J., *Las Relaciones Topográficas de Felipe II. Índices, Fuentes y Bibliografía* (El Escorial: Real Centro Universitario Escorial María Cristina, 2007), at http://www.rcumariacristina.com/ficheros/JavierCampos [accessed 4 May 2013].

Canguilhem, G., 'La Monstruosité et le Monstrueux', in *La Connaissance de la Vie* (Paris: Vrin, 1980), pp. 178–9.

—, *The Normal and the Pathological*, trans. C. R. Fawcett and R. S. Cohen (New York: Zone Books, 1989).

Canterla, C., *Mala Noche. El cuerpo, la política y la irracionalidad en el siglo XVIII* (Seville: Fundación José Manuel Lara, 2009).

Capuron, J., *Tratado de las Enfermedades de las Mugeres desde la edad de la pubertad hasta la crítica inclusive*, 2 vols (Madrid: Imprenta Calle de la Greda, 1821).

Cardoso, I., *Philosophia libera in septem libros distributa* (Venice: Bertanorum sumptibus, 1673), at http://books.google.co.uk/books?id=c0BfJaN5mA4C&printsec=frontcover&hl=es&source=gbs_ge_summary_r&cad=0#v=onepage&q&f=false [accessed 22 August 2013]

Carranza, A., *De Partu Naturali et Legitimo* (Cologne: Jacobi Stor, 1630).

Carrasco, R., and B. Vicent, 'Amours et marriage chez les morisques au Xv^e siècle', in A. Redondo (ed.), *Amours Légitimes et Illégitimes en Espagne (XVIe–XVIIe siècles)* (Paris: Publications de la Sorbonne, 1985), pp. 133–50.

Cartagena-Calderón, J., '"Él es tan rara persona". Sobre cortesanos, lindos, sodomitas y otras masculinidades nefandas en la España de la temprana Edad Moderna', in M. J. Delgado and A. Saint-Saëns (eds), *Lesbianism and Homosexuality in Early Modern Spain* (New Orleans, LA: University Press of the South, 2000), pp. 139–76.

Castel, R., *Les métamorphoses de la question sociale: une chronique du salariat* (Paris: Fayard, 1995).

Castro, E. Rodrigo de, *Tractatus de natura muliebri* (Hamburg: Sande, 1608), at http://search.ugent.be/meercat/x/bkt01?q=900000147674 [accessed 22 August 2013].

Cátedra, M., 'Sobre la ambigüedad: el caso de Paula Barbada', in E. Crespo and C. Soldevilla (eds), *La Constitución Social de la Subjetividad* (Madrid: Los Libros de La Catarata, 2001), pp. 131–44.

Chinchilla, A., *Anales históricos de la medicina en general y biográficos-bibliográficos de la española en particular: Historia de la Medicina Española*, 4 vols (New York and London: Johnson Reprint Corporation, 1967).

Clementi, H., *La Frontera en América. Una clave interpretativa de la historia americana* (Buenos Aires: Leviatán, 1985).

Cleminson, R., and R. Medina Doménech, '¿Mujer u hombre? Hermafroditismo, tecnologías médicas e identificación del sexo en España, 1860–1925', *Dynamis*, 24 (2004), pp. 53–91.

Cleminson, R., and F. Vázquez García, *Hermaphroditism, Medical Science and Sexual Identity in Spain, 1850–1960* (Cardiff: University of Wales Press, 2009).

—, and —, 'The Hermaphrodite, Fecundity and Military Efficiency: Dangerous Subjects in the Emerging Liberal Order of Nineteenth-Century Spain', in K. Fisher and S. Toulalan (eds), *Bodies, Sex and Desire from the Renaissance to the Present* (Basingstoke: Palgrave Macmillan, 2011), pp. 70–86.

Cochrane, E. W., *Historians and Historiography in the Italian Renaissance* (Chicago, IL: University of Chicago Press, 1981).

Comte Masía, M. (ed.), *Casamiento entre dos damas* (1849; Madrid: Imp. de D. José Mª Marés, 2000).

Cordoba, P., 'L'homme enceint de Grenade. Contribution à un dossier d'histoire culturelle', *Mélanges de la Casa de Velázquez*, 23 (1987), pp. 307–30.

Correa Calderón, E., *Baltasar Gracián. Su vida y su obra* (Madrid: Gredos, 1961).

Costa, A. da, *A mulher em Portugal* (Lisbon: Typ. da Companhia Nacional Editora, 1892).

Covarrubias, S., *Tesoro de la Lengua Castellana o Española* (1611; Madrid: Turner, 1979).

Coyne Kelly, K., *Performing Virginity and Testing Chastity in the Middle Ages* (London and New York: Routledge, 2000).

Crawford, K., 'The Good, the Bad, and the Textual: Approaches to the Study of the Body and Sexuality, 1500–1750', in S. Toulalan and K. Fisher (eds), *The Routledge History of Sex and the Body 1500 to the Present* (London and New York: Routledge, 2013), pp. 23–37.

Dany, M., 'Observación que puede servir para la historia del hermafrodismo', *Gaceta Médica de Madrid*, 1 (1835), p. 151.

Darmon, P., *Trial by Impotence* (London: Chatto & Windus, 1985).

Daston, L., 'Marvelous Facts and Miraculous Evidence in Early Modern Europe', *Critical Inquiry*, 18 (1991), pp. 93–124.

—, and K. Park, 'Hermaphrodites in Renaissance France', *Critical Matrix: Princeton Working Papers in Women's Studies*, 1:5 (1985), pp. 1–19.

—, and —, *Wonders and the Order of Nature, 1150–1750* (New York: Zone Books, 1988).

—, and —, 'The Hermaphrodite and the Orders of Nature: Sexual Ambiguity in Early Modern France', *GLQ: A Journal of Gay and Lesbian Studies*, 1:4 (1995), pp. 420–5.

—, and —, 'The Hermaphrodite and the Orders of Nature: Sexual Ambiguity in Early Modern France', in L. Fradenburg and C. Freccero (eds), *Premodern Sexualities* (New York and London: Routledge, 1996), pp. 117–36.

Daudin, H., *Cuvier et Lamarck. Les classes zoologiques et l'idée de série animale*, 2 vols (Paris: Alcan, 1926–7).

—, *De Linné a Lamarck. Méthodes de la classification et idée de série en Botanique et en Zoologie (1740–1790)* (Paris: Alcan, 1926–7).

De Arriaga, J., *Piscator Murciano. Con un agregado de prodigios, cosas no comunes y fuera de el estado natural, que han sucedido, y dignas de que se sepan, como haberse buelto muchas mujeres hombres* (Madrid, 1746).

De Centellas, L., *Coplas de Don Luis de Centellas sobre la Piedra Filosofal* (c. 1552), in *Cinco Tratados Españoles de Alquimia* (Madrid: Tecnos, 1987).

De Fuentelapeña, A., *El Ente Dilucidado. Tratado de Monstruos y Fantasmas* (1676; Madrid: Editora Nacional, 1978).

De Fuentes, A., *Summa de Philosophía Natural* (Seville: J. León, 1547).

De Granada, L., *Introducción al Símbolo de la Fe*, in *Obras*, vol. 1 (1583; Madrid: BAE, 1944), pp. 81–633.

De Jaucourt, C., 'Hermaphrodite', in *Encyclopédie de Diderot et D'Alembert*, vol. 15 (1755; Paris: Ed. de Franco Maria Ricci, 1978), fols H27–H28.

De la Cerda, J., *Libro intitulado vida política de todos los estados de mujeres* (Alcalá de Henares: Juan Gracián, 1599).

De la Flor, F., 'La "puella pilosa". Representaciones de la alteridad femenina (de Sánchez Cotán a José de Ribera, pasando por Sebastián de Covarrubias)', in *La Península Metafísica. Arte, literatura y Pensamiento en la España de la Contrarreforma* (Madrid: Biblioteca Nueva, 1999), pp. 267–305.

De la Pascua Sánchez, M. J., '¿Hombres vueltos al revés? Una historia sobre la construcción de la identidad sexual en el siglo XVIII', in M. J. de la Pascua Sánchez, M. del Rosario García Doncel and G. Espigado (eds), *Mujer y deseo* (Cadiz: Universidad de Cádiz, 2004), pp. 431–44.

De León, A., *Libro Primero de Anatomía: recopilaciones y examen general* (Baeza: Juan Baptista de Montoya, 1590).

De Lletor Castroverde, J., *Repertorio Médico Extranjero*, vol. 5 (Madrid: Imprenta Real, 1835).

De Pineda, J., *Treinta y Cinco Diálogos Familiares de Agricultura Cristiana* (Salamanca: Pedro de Adurça y Diego López, 1589).

De Santa María, J., *República y policía Christiana* (Lisbon, 1621).

De Torquemada, A., *Jardín de Flores Curiosas* (1570; Madrid: Sociedad de Bibliófilos Españoles, 1943).

De Torreblanca y Villalpando, F., *Epithomes delictorum sive de Magia* (Lyon: Juan Antonio Huguet, 1678).

De Viguera, B., *La Fisiología y Patología de la Muger o sea historia analítica de su constitución física y moral, de sus atribuciones y fenómenos sexuales y de todas sus enfermedades*, 2 vols (Madrid: Imprenta de Ortega y Compañía, 1827).

De Vun, L., 'The Jesus Hermaphrodite: Science and Sex Difference in Premodern Europe', *Journal of the History of Ideas*, 69:2 (2008), pp. 193–218.

Dean, M., *Governmentality: Power and Rule in Modern Society* (London: Sage, 1999).

Dean-Jones, L., *Women's Bodies and Classical Greek Science* (Oxford: Clarendon Press, 1994).

Del Río, Martín, *La Magia Demoníaca. Libro II de las Disquisiciones Mágicas* (1599–1600; Madrid: Hiperión, 1991).

Delcourt, M., *Hermafrodita. Mitos y Ritos de la Sexualidad en la Antigüedad Clásica* (Barcelona: Seix Barral, 1970).

Deleuze, G., 'What is a Dispositif?', in T. J. Armstrong (ed. and trans.), *Michel Foucault: Philosopher* (Hemel Hempstead: Harvester Wheatsheaf, 1992), pp. 159–66.

—, *The Fold: Leibniz and the Baroque*, trans. T. Conley (Minneapolis, MN: University of Minnesota Press, 1993).

Delpech, F., '"Muger hay en la guerra": Remarques sur l'exemplaire et curieuse carrière d'une guerrière travestie, Juliana de los Cobos', in A. Redondo (ed.), *Relations entre hommes et femmes en Espagne aux XVIe et XVIIe siècles* (Paris: Presses de l'Université de la Sorbonne, 1995), pp. 53–65.

Del Río Parra, E., *Una era de monstruos. Representaciones de lo deforme en el Siglo de Oro español* (Madrid: Iberoamericana, 2003).

Díaz Gito, M. A., 'El Poema "Corsica" de J. Cristóbal Calvete de Estrella (y otros dos poemas latinos)' (unpublished doctoral thesis, University of Cadiz, 1990).

Díez Fernández, J. I., *La poesía erótica de los Siglos de Oro* (Madrid: Ediciones del Laberinto, 2003).

—, *Tres discursos de mujeres. (Poética y hermenéutica cervantinas)* (Alcalá de Henares: Centro de Estudios Cervantinos, 2004).

Donato, M., *De medica historia mirabili* (Venice, 1588).

Donoghue, E., 'Imagined More than Women: Lesbians as Hermaphrodites, 1671–1766', *Women's History Review*, 2:2 (1993), pp. 199–216.

Dekker, R. M., and L. C. Van de Pol, *The Tradition of Female Transvestism in Early Modern Europe* (London: Macmillan, 1989).

Dreger, A. D., *Hermaphrodites and the Medical Invention of Sex* (Cambridge, MA and London: Harvard University Press, 1998).

Eiximenis, F. de, *Carro de las Donas, trata de la vida y muerte del hombre cristiano* (Valladolid: Juan de Villaquirán, 1542).

Efron, N. J., 'Nature, Human Nature, and Jewish Nature in Early Modern Europe', *Science in Context*, 15:1 (2002), pp. 29–49.

Eliade, M., *Mefistófeles y el Andrógino* (Barcelona: Labor, 1984).

Elias, N., *The Court Society*, trans. E. Jephcott (Oxford: Blackwell, 1983).

—, *The Civilizing Process: Sociogenetic and Psychogenetic Investigations*, trans. E. Jephcott (Malden, Oxford and Carlton: Blackwell, 2000).

Enguix, B., 'Cuerpo y transgresión: de Helena de Céspedes a Lady Gaga', *Revista Latinoamericana de Estudios sobre Cuerpos, Emociones y Sociedad*, 5:3 (2011), pp. 25–38.

Epstein, J., 'Either/Or-Neither/Both: Sexual Ambiguity and the Ideology of Gender', *Genders*, 7 (1990), pp. 99–142.

Escalonilla López, R. A., 'Función del Travestismo en las Comedias de Don Pedro Calderón de la Barca' (unpublished doctoral thesis, Universidad Politécnica de Madrid, 1998).

Escamilla, M., 'A propos d'un dossier inquisitorial des environs de 1590: les étranges amours d'un hermaphrodite', in A. Redondo (ed.), *Amours légitimes, Amours illégitimes en Espagne (XVIe–XVIIe Siècles)* (Paris: Pub. de la Sorbonne, 1985), pp. 167–82.

Esdaile, C. J., *The Spanish Army in the Peninsular War* (Manchester and New York: Manchester University Press, 1988).

Eslava Galán, J., 'Introducción General', in *Cinco Tratados Españoles de Alquimia* (Madrid: Tecnos, 1987), pp. 13–47.

—, 'La Alquimia Española del Siglo de Oro', in *Cinco Tratados Españoles de Alquimia* (Madrid: Tecnos, 1987), pp. 114–16.

—, *Historia secreta del sexo en España* (Barcelona: Planeta, 1992).

Espósito, R., *Bíos. Biopolítica y Filosofía* (Buenos Aires: Amorrortu, 2006).

Ettinghausen, H., 'Sexo y Violencia: noticias sensacionalistas en la prensa española del siglo XVII', *Edad de Oro*, 12 (1993), pp. 95–107.

— (ed.), *Noticias del Siglo XVII: Relaciones Españolas de Sucesos Naturales y Sobrenaturales* (Barcelona: Puvill Libros, 1995).

Farley, J., *Gametes and Spores: Ideas about Sexual Reproduction 1750–1914* (Baltimore, MD: Johns Hopkins University Press, 1982).

Fausto-Sterling, A., *Sexing the Body: Gender Politics and the Construction of Sexuality* (New York: Basic Books, 2000).

Feijóo, B. J., 'Defensa de las Mujeres', in *Teatro Crítico Universal*, vol. 1 (1726; Madrid: Imp. de Don Joaquín Ibarra, 1778), pp. 325–98.

—, 'La Doctrina Hipocrática no debe tomarse por norma de Medicina', in *Teatro Crítico Universal*, vol. 8 (1739; Madrid: Imp. de Don Joaquín Ibarra, 1779), pp. 328–39.

—, 'Respuesta a la consulta sobre el infante monstruoso de dos cabezas, dos cuellos, cuatro manos ... que salió a luz en Medina Sidonia el 24 de febrero del año 1736', in *Cartas Eruditas y Curiosas*, vol. 1 (1742; Madrid: Imprenta Real de la Gaceta, 1777), pp. 78–100.

—, 'Reflexiones filosóficas, con ocasión de una criatura humana hallada poco ha en el vientre de una cabra', in *Cartas Eruditas y Curiosas*, vol. 3 (1750; Madrid: Imprenta Real de la Gaceta, 1774), pp. 327–52.

Fernández del Valle, J., *Cirugía forense, general y particular*, vol. 1 (Madrid: Imprenta de Aznar, 1796).

—, *Cirugía forense, general y particular*, vol. 3 (Madrid: Imprenta de Aznar, 1797).

Fernández Navarrete, P., *Conservación de Monarquías y Discursos Políticos* (1626; Madrid: Instituto de Estudios Fiscales, Ministerio de Hacienda, 1982).

Fernández-Santamaría, J. A., *Razón de Estado y Política en el Pensamiento Español del Barroco (1595–1640)* (Madrid: Centro de Estudios Constitucionales, 1986).

Feros, A., *El Duque de Lerma. Realeza y privanza en la España de Felipe III* (Madrid: Marcial Pons, 2009).

Ferrer y Garcés, R., *Tratado de Medicina Legal* (Barcelona: Imprenta de P. Riera, 1847).

Fletcher, R., *Moorish Spain* (London: Phoenix, 1998).

Foderé, F. E., *Las leyes ilustradas por las ciencias físicas ó tratado de medicina legal y de higiene pública*, 8 vols (1798; Madrid: Imp. de la Administración del Real Arbitrio de Beneficencia, 1801–3).

—, *Les lois éclairées par les sciences physiques, ou traité de médecine légale et d'hygiène publique* (Paris: chez Croullebois et chez Deterville, 1813).

Folch Jou, G., and M. S. Muñoz Calvo, 'Un pretendido caso de hermafroditismo en el siglo XVI', *Boletín de la Sociedad de Historia de la Farmacia*, 93 (1973), pp. 20–33.

Fontes, J. L. Inglês, J. Bastos Serra and M. F. Andrade, *Inventário dos Fundos Monástico-Conventuais da Biblioteca Pública de Évora* (Lisbon and Évora: Edições Colibri/Universidade de Évora, 2010).

Foucault, M., *El orden del discurso*, trans. A. González Troyano (1970; Barcelona: Tusquets, 2002).

Foucault, M., *Herculine Barbin dite Alexina B.* (Paris: Gallimard, 1978).

—, *Herculine Barbin, Being the Recently Discovered Memoirs of a Nineteenth-Century French Hermaphrodite*, trans. R. McDougall (New York: Pantheon Books, 1980).

—, *La verdad y las formas jurídicas* (Barcelona: Gedisa, 1980).

—, *The Birth of the Clinic: An Archaeology of Medical Perception*, trans. A. M. Sheridan (London: Routledge, 1989).

—, 'La vida de los hombres infames', in *La Vida de los Hombres Infames, Ensayos sobre desviación y dominación* (Madrid: La Piqueta, 1990), pp. 175–202.

—, *The History of Sexuality, Volume 1: An Introduction* (Harmondsworth: Penguin, 1990).

—, 'Chronologie', in *Dits et Écrits 1954–1988*, vol. 1 (Paris: Gallimard, 1994), p. 54.

—, 'Le vrai sexe', in *Dits et Écrits 1954–1988*, vol. 4 (Paris: Gallimard, 1994), pp. 115–23.

—, *'Il Faut Défendre la Société'. Cours au Collège de France, 1976* (Paris: Seuil, 1997).

—, *Les Anormaux. Cours au Collège de France. 1974–1975* (Paris: Gallimard-Le Seuil, 1999).

—, *Securité, Territoire, Population. Cours au Collège de France 1977–1978* (Paris: Gallimard-Le Seuil, 2004).

Fragoso, J., *Erotemas Chirúrgicos* (Madrid: Sebastián Yánez, 1570).

—, *Cirugía Universal* (1581; Madrid: Viuda de Alonso Martín, 1627).

Fraile, P., *La Otra Ciudad del Rey. Ciencia de la Policía y organización urbana en España* (Madrid: Celeste Ediciones, 1997).

Franco, G. dos Reis, *Elysius iucundarum quaestionum campus* (Brussels: Typis & sumptibus Francisci Vivien, 1661).

Friedenwald, H., *Jewish Luminaries in Medical History* (Baltimore, MD: Johns Hopkins Press, 1946).

Friesen, I. E., *The Female Crucifix: Images of St. Wilgefortis since the Middle Ages* (Waterloo: Wilfrid Laurier University Press, 2001).

Front, D., 'The Expurgation of Medical Books in Sixteenth-Century Spain', *Bulletin of the History of Medicine*, 75:2 (2001), pp. 290–6.

Fuchs, B., 'Border Crossings: Transvestism and Passing in "Don Quijote"', *Cervantes: Bulletin of the Cervantes Society of America*, 16:2 (1996), pp. 4–28.

García-Ballester, L., *Los Moriscos y la Medicina. Un capítulo de la medicina y la ciencia marginadas en la España del siglo XVI* (Barcelona: Labor, 1984).

—, 'The Circulation and Use of Medical Manuscripts in Arabic in Sixteenth-Century Spain', in J. Arrizabalaga, M. Cabré, L. Cifuentes and F. Salmón (eds), *Galen and Galenism: Theory and Medical Practice from Antiquity to the European Renaissance* (Aldershot and Burlington, VT: Ashgate, 2002), pp. 183–99.

García Carrero, P., *Disputationes Medicae Super Libros Galeni de Locis Affectis et de Aliis Morbis ab eo Relictis* (Alcalá de Henares: J. Sánchez Crespo, 1605).

García de Enterría, M. Cruz, H. Ettinghausen, V. Infantes and A. Redondo (eds), *Las Relaciones de sucesos en España (1500–1750). Actas del Primer Coloquio Internacional (Alcalá de Henares, 8, 9 y 10 de junio de 1995)* (Alcalá de Henares and Paris: Pub. Universidad Alcalá de Henares/Pub. Sorbonne, 1996).

García Font, J., *Historia de la Alquimia en España* (Madrid: Editora Nacional, 1976).

Garza Carvajal, F., *Quemando Mariposas. Sodomía e Imperio en Andalucía y México, siglos XVI–XVII* (Barcelona: Laertes, 2002).

—, *Las Cañitas. Un proceso por lesbianismo a principios del XVII* (Palencia: Simancas Ediciones, 2012).

Gilbert, R., *Early Modern Hermaphrodites* (Basingstoke: Palgrave, 2002).

Giraldo, C., 'Una hija de Hermes y Afrodita en un expediente de 1803' (2008), at http://www.incertidumbre.org/index.php?option=com_content&task=view&id=85&Itemid=55 [accessed 15 May 2013].

Gómez de Huerta, J., *Historia Natural de Cayo Plinio Segundo* (Madrid: Iván González, 1629).

González Cañal, R., 'El lujo y la ociosidad durante la privanza de Olivares: Bartolomé Jiménez Patón y la polémica sobre el guardainfante y las guedejas', *Criticón*, 53 (1991), pp. 71–96.

González Gómez, O., 'Los discursos sobre la sodomía en las Historias de la conquista de América' (2007), at http://posgradoestudioslatinoamericanos.blogspot.com.es/2009_12_01_archive.html [accessed 11 April 2013].

González Rovira, J., 'Imaginativa y nacimientos prodigiosos en algunos textos del Barroco', *Criticón*, 69 (1997), pp. 21–31.

Goulemot, J. M., 'Fureurs utérines', *Dix-Huitième Siècle*, 12 (1980), pp. 97–111.

Graille, P., *Le troisième sexe. Être hermafrodite aux XVIIe et XVIIIe siècles* (Paris: Les éditions arkhê, 2011).

Granjel, L. S., *Médicos Españoles* (Salamanca: Universidad de Salamanca, 1967).

—, *La Medicina Española del siglo XVIII* (Salamanca: Universidad de Salamanca, 1979).

—, *La Medicina Española Renacentista* (Salamanca: Universidad de Salamanca, 1980).

—, 'El Médico Andrés Laguna', in J. L. García Hourcade and J. M. Moreno Yuste (eds), *Andrés Laguna. Humanismo, Ciencia y Política en la Europa Renacentista* (Valladolid: Junta de Castilla y León, Consejería de Educación y Cultura, 2001), pp. 11–16.

Granjel, M., *Pedro Felipe Monlau y la Higiene Española del siglo XIX* (Salamanca: Universidad de Salamanca, 1983).

Greenblatt, S., *Renaissance Self-Fashioning: From More to Shakespeare* (Chicago, IL: University of Chicago Press, 1980).

—, *Shakespearean Negotiations: The Circulation of Social Energy in Renaissance England* (Oxford: Clarendon Press, 1988).

Grimson, A. (ed.), *Fronteras, Naciones e Identidades. La periferia como centro* (Buenos Aires: CICCUS-La Crujía, 2000).

Guicciardi, J. P., 'L'hermaphrodite et le prolétaire', *Dix-Huitième Siècle*, 12 (1980), pp. 49–78.

Gutiérrez de Torres, A., *El Sumario de las Maravillosas y Espantables Cosas que en el Mundo han acontecido* (1524; Madrid: Real Academia de la Lengua Española, 1952).

Gutiérrez Nieto, A., 'El pensamiento económico político y social de los arbitristas', in M. Menéndez Pidal (ed.), *Historia de España. El Siglo del Quijote*, vol. 1 (Madrid: Espasa Calpe, 1986), pp. 233–351.

Hagner, M., 'Utilidad científica y exhibición pública de monstruosidades en la época de la Ilustración', in A. Lafuente and J. Moscoso (eds), *Monstruos y Seres Imaginarios en la Biblioteca Nacional* (Madrid: Ministerio de Educación y Cultura, Biblioteca Nacional, 2000), pp. 105–28.

Haliczer, S., 'The First Holocaust: The Inquistion and the Converted Jews of Spain and Portugal', in *Inquisition and Society in Early Modern Europe* (London and Sydney: Croom Helm, 1987), pp. 7–18.

Harvey, K., 'The Substance of Sexual Difference: Change and Persistence in Representations of the Body in Eighteenth-Century England', *Gender and History*, 14:2 (2002), pp. 202–23.

Herculano, A., *Historia da origem e estabelecimento da Inquisição em Portugal*, ed. D. Lopes, 3 vols (Rio de Janeiro, São Paulo, Belo Horizonte and Lisbon: Editoria Paulo de Azevedo/ Livraria Bertrand, 1925).

Herdt, G. (ed.), *Third Sex, Third Gender: Beyond Sexual Dimorphism in Culture and History* (New York: Zone Books, 2000).

Hernández Morejón, A., *Historia bibliográfica de la medicina española*, vol. 5 (Madrid: Imprenta de la Viuda de Jordán é Hijos, 1847).

Hervás y Panduro, L., *Historia de la Vida del Hombre o idea del Universo*, vol. 1 (Madrid: Aznar, 1789).

Hespanha, A. M., *Vísperas del Leviatán. Instituciones y poder político (Portugal, siglo XVII)* (Madrid: Taurus, 1989).

Higgs, D., 'Tales of Two Carmelites: Inquisitorial Narratives from Portugal and Brazil', in P. Sigal (ed.), *Infamous Desire: Male Homosexuality in Colonial Latin America* (Chicago, IL: Chicago University Press, 2003), pp. 152–67.

Homero Arjona, J., 'El disfraz varonil en Lope de Vega', *Bulletin Hispanique*, 39 (1937), pp. 120–45.

Hotchkiss, V., *Clothes Make the Man: Female Cross Dressing in Medieval Europe* (London: Routledge, 1996).

Huarte de San Juan, J., *Examen de Ingenios para las Ciencias* (1575; Madrid: Editora Nacional, 1977).

Huertas, R., *Orfila, Saber y Poder Médico* (Madrid: CSIC, 1988).

Hurteau, P., 'Catholic Moral Discourse on Male Sodomy and Masturbation in the Seventeenth and Eighteenth Century', *Journal of the History of Sexuality*, 4:1 (1993), pp. 1–26.

Hull, I. V., *Sexuality, State and Civil Society in Germany, 1700–1815* (Ithaca, NY: Cornell University Press, 1996).

Hurtado de Mendoza, A., 'Hermafrodismo', in *Suplemento al Diccionario de Medicina y Cirugía del Profesor D. Antonio Ballano*, vol. 3 (Madrid: Imprenta de Brugada, 1823), p. 1135.

Hurtado de Mendoza, M., *Instituciones de Medicina*, vol. 1 (Madrid: Imp. Sánchez, 1839).

—, *Vocabulario Médico-Quirúrgico o Diccionario de Medicina y Cirugía* (Madrid: Boix Editor, 1840).

Hutcheson, G. S., 'The Sodomitic Moor: Queerness in the Narrative of *Reconquista*', in G. Burger and S. F. Kruger (eds), *Queering the Middle Ages* (Minneapolis, MN and London: University of Minnesota Press, 2001), pp. 99–122.

Iglesias Aparicio, P., *Construcción del sexo y género desde la Antigüedad al siglo XIX* (Vigo: Universidad de Vigo, 2007), at http://webs.uvigo.es/pmayobre/textos/pilar_iglesias_aparicio/tesis_doctoral/cap2_construccion_de_sexo_y_genero_desde_%20la_edad_media.doc [accessed 14 April 2013].

Inamoto, K., 'La mujer vestida de hombre en el teatro de Cervantes', *Cervantes: Bulletin of the Cervantes Society of America*, 12:2 (1992), pp. 137–43.

Jacob, F., *The Logic of Life: A History of Heredity. The Possible and the Actual*, trans. B. E. Spillman (London: Penguin, 1989).

Jacquart, D., 'Influence de la médicine arabe en Occident médiéval', in R. Rashed (ed.), *Histoire des Sciences Arabes, vol. III : Technologie, alchimie et sciences de la vie* (Paris: Éditions du Seuil, 1997), pp. 213–32.

—, and C. Thomasset, *Sexualidad y Saber Médico en la Edad Media* (Barcelona: Labor, 1989).

Jiménez Monteserín, M., *Sexo y bien común. Notas para la historia de la prostitución en España* (Cuenca: Ayuntamiento de Cuenca, Instituto 'Juan de Valdés', 1994).

Jovellanos, G. M., 'Sátira primera a Arnesto', in *Obras Completas, vol. I: Obras literarias* (1786; Oviedo: Centro de Estudios del Siglo XVIII, Ilustre Ayuntamiento de Gijón, 1984), pp. 220–7.

—, 'Sátira segunda a Arnesto. Sobre la mala educación de la nobleza', in *Obras Completas, vol. I: Obras literarias* (1787; Oviedo: Centro de Estudios del Siglo XVIII, Ilustre Ayuntamiento de Gijón, 1984), pp. 227–44.

Kagan, R., and A. Dyer, *Inquisitorial Inquiries: Brief Lives of Secret Jews and Other Heretics* (Baltimore, MD: Johns Hopkins University Press, 2004).

Kamen, H., *Cambio cultural en la sociedad del Siglo de Oro. Cataluña y Castilla, siglos XVI–XVII* (Madrid: Siglo XXI, 1998).

—, *The Disinherited: The Exiles Who Created Spanish Culture* (London: Penguin, 2008).

Kappler, C., *Monstruos, Demonios y Maravillas a fines de la Edad Media* (Madrid: Akal, 1986).

Knibiehler, Y., and C. Fouqué, *La Femme et les Médecins* (Paris: Hachette, 1983).

Koyré, A., *Mystiques, spirituels, alchimistes du XVIe siècle allemand* (Paris: Gallimard, 1971).

Lacarra Lanz, E., 'Homoerotismo femenino en los discursos normativos medievales', in A. Chas Aguión and C. Tato García (eds), *Siempre soy quien ser solía: Estudios de literatura española medieval en homenaje a Carmen Parrilla* (A Coruña: Universidade da Coruña, 2009), pp. 205–28.

Lalinde Abadía, J., 'La indumentaria como símbolo de discriminación jurídico social', *Anuario de Historia del Derecho Español*, 53 (1986), pp. 583–601.

Laqueur, T., *Making Sex: Body and Gender from the Greeks to Freud* (Cambridge, MA and London: Harvard University Press, 1990).

—, 'Sex in the Flesh', *Isis*, 94:2 (2003), pp. 300–6.

Le Goff, J., *The Medieval Imagination*, trans. A. Goldhammer (Chicago, IL and London: University of Chicago Press, 1992).

Leão, D. Nunez do, *Descripção do Reino de Portugal* (Lisbon: Iorge Rodriguez, 1610).

Lehfeldt, E. A., 'Ideal Men: Masculinity and Decline in Seventeenth-Century Spain', *Renaissance Quarterly*, 61:2 (2008), pp. 463–94.

Lemos, M., *Amato Lusitano: A sua vida e a sua obra* (Oporto: Typ. da Encyclopedia Portugueza Ilustrada, 1907).

León Pinelo, A. de, 'Velos antiguos y modernos en los rostros de las mujeres. Sus conveniencias y daños' (1639), *Lemir*, 13 (2009), pp. 235–388.

Libis, J., *El Mito del Andrógino* (Madrid: Siruela, 2001).

Lima, Castro Pires de, *A Mulher vestida de Homem (Contribuição para o estudo do romance 'A Donzela que vai à Guerra')* (Coimbra: Fundação Nacional para a Alegria no Trabalho-Gabinete de Etnografia, 1958).

Lima, J. A. Pires de, 'Contribuïção para a história da Teratologia portuguesa. Monstros duplos autositários', *Arquivo de Anatomia e Antropologia*, 18 (1937), pp. 53–60.

—, 'Hermafroditismo e Inter-Sexualidade', *A Medicina Contemporânea*, 44 (1939), pp. 473–8.

Lluch, E., 'El Cameralismo en España', in E. Fuentes Quintana (ed.), *Economía y Economistas Españoles. Tomo III. La Ilustración* (Barcelona: Galaxia Gutenberg, 2000), pp. 721–8.

Long, K. P., *Hermaphrodites in Renaissance Europe* (Aldershot: Ashgate, 2006).

Lobera de Avila, L., *Libro de Regimiento de la Salud y de la esterilidad de los hombres y mujeres* (Valladolid: Sebastián Martínez, 1551).

López Mateos, R., *Pensamientos sobre la razón de las leyes* (Madrid: Imp. Gómez Fontenebro y Compañía, 1810).

López Piñero, J. M., T. F. Glick, V. Navarro Brótons and E. Portela Marco, *Diccionario Histórico de la Ciencia Moderna en España*, 2 vols (Barcelona: Península, 1983).

Lorenzo Vélez, A., 'El motivo de "La Mujer Disfrazada de Varón" en la tradición oral moderna', *Revista de Folklore*, 17a:194 (1997), pp. 39–59.

Lorenzo Zambrano, M. R. P. M., 'Si es posible el concurso carnal del demonio con criatura humana y en este caso habiendo prole, si es capaz de bautismo', *Memorias Académicas de la Real Sociedad de Medicina y demás Ciencias de Sevilla*, 9 (1790), pp. 409–22.

Lusitani, A., *Curationum medicinalium* (Lugduni [Lyon]: Gulielmum Rouillium, 1567).

Lusitani, Castro R., *De universa muliebrium morborum medicina*, 3rd edn (Hamburg: Ex bibliopolio Frobbeniano, 1628).

Lusitani, Z., *Operum Tomos Primus in quo de Medicorum Principum Historia* (Lyon: Joanni Antonii Huguetan & Marci Antonii Ravaud, 1649).

Maganto Pavón, E., 'La intervención del Dr. Francisco Díaz en el proceso inquisitorial contra Elena/o de Céspedes, una cirujana transexual condenada por la Inquisición de Toledo en 1587', *Historia de la Urología Española*, 60:8 (2007), pp. 873–86.

—, *El proceso inquisitorial contra Elena/o de Céspedes (1587–1588) (Biografía de una cirujana transexual del siglo XVI)* (Madrid: Método Gráfico, 2007).

Margouliès, G., 'Deux poèmes sur la jeune fille partie à la guerre por remplacer son père', *Revue de Littérature Comparée*, 8 (1928), pp. 304–9.

Mantecón Movellán, T., 'Los mocitos de Galindo: sexualidad *contra natura*, culturas proscritas y control social en la edad moderna', in T. Mantecón Movellán (ed.), *Bajtín y la historia de la cultura popular* (Santander: Universidad de Cantabria, 2008), pp. 209–40.

Marañón, G., *Las Ideas Biológicas del Padre Feijóo* (Madrid: Espasa Calpe, 1954).

Maravall, J. A., *La Cultura del Barroco* (Barcelona: Ariel, 1983).

—, *Teoría del Estado en España en el siglo XVII* (Madrid: Centro de Estudios Constitucionales, 1997).

Marc, C. C. H., 'Hermaphrodite', in *Dictionnaire des Sciences Médicales par une Société de Médecins et de Chirurgiens, Vol. XXI: HEM–HUM* (Paris: C. L. F. Panckoucke, 1817), pp. 86–121.

Marchetti, V. 'Propositions de règlement juridique d'une troisième sexualité: Lorenzo Matheu y Sanz et les hermaphrodites', in J. Poumarede and J. P. Royer (eds), *Droit, histoire et sexualité* (Lille: Université de Lille et Université de Toulouse, 1987), pp. 131–43.

—, 'La discussione secentesca sui diritti dei bissesuali', in S. R. Ghibaudi and F. Barcia (eds), *Studi Politici in Onore di Luigi Firpo*, 4 vols (Turin: F. Angeli, 1990), vol. 2, pp. 463–74.

—, *L'invenzione della bissesualità. Discussioni tra teologi, medici e giuristi del XVII secolo sull'ambiguità dei corpi e delle anime* (Milan: Mondadori, 2001).

Marqués de Armas, P., 'El Monstruo Humano' (2002), at http://www.habanaelegante.com/Summer2002/Panoptico.html [accessed 15 February 2013].

Martín, A. L., 'Sodomitas, putos, doncellos y maricotes en algunos textos de Quevedo', *La Perinola*, 12 (2008), pp. 107–22.

Martín Rodríguez, M., *Pensamiento Económico Español sobre la Población* (Madrid: Ediciones Pirámide, 1984).

Martínez, M., *Anatomía Completa del Hombre* (1728; Madrid: Imp. de la Viuda de Manuel Fernández, 1764).

—, *Observatio Rara de Corde in Monstroso Infantulo ubi obiter et noviter de motu cordis et sangunis agitur* (Madrid: Francisco Rodríguez, 1750).

Martínez, R., 'Mari(c)ones, travestis y embrujados. La heterodoxia del varón como recurso cómico en el teatro breve del Barroco', *Anagórisis*, 3 (2011), pp. 9–37.

Martínez Pérez, J., 'Medicina, Liberalismo y Legislación: Ramón López Mateos (1771–1814) y sus *Pensamientos sobre la razón de las leyes*', *Asclepio. Revista de Historia de la Medicina y de la Ciencia*, 40:2 (1988), pp. 209–46.

—, 'Sexualidad y Orden Social: la visión médica de la España del primer tercio del siglo XIX', *Asclepio*, 42:2 (1990), pp. 119–35.

—, 'Colegios de Cirugía y Medicina Legal: una expresión de los procesos de intercambio entre fuerzas armadas y sociedad a finales del siglo XVIII', in E. Balaguer and E. Giménez (eds), *Ejército, Ciencia y Sociedad en la España del Antiguo Régimen* (Alicante: Instituto de Cultura Juan Gil-Albert, 1995), pp. 493–511.

Mata, P., *Vademecum de Medicina y Cirugía Legal*, vol. 1 (Madrid: Imprenta Calle de Padilla, 1844).

—, *Tratado de Medicina y Cirugía Legal Teórica y Práctica* (Madrid: Carlos Bailly Baillière, 1846).

—, *Tratado teórico-práctico de Medicina Legal y Toxicología* (Madrid: Bailly-Baillière e Hijos, 1903).

Matheu i Sanz, L., *Tractatus de re criminali* (1686; Lyon: Ioann. Posvel et Claudium Rigaud, 1686).

Mayer, A. J., *The Persistence of the Old Regime: Europe to the Great War* (New York: Pantheon Books, 1981).

McKendrick, M., *Woman and Society in the Spanish Drama of the Golden Age: A Study of the 'Mujer Varonil'* (Cambridge: Cambridge University Press, 1974).

McVaugh, M. R., *Medicine before the Plague: Practitioners and their Patients in the Crown of Aragon, 1283–1345* (Cambridge: Cambridge University Press, 1993).

Mead, G. H., 'La genesis del *self* y el control social', *Revista Española de Investigaciones Sociológicas*, 55 (1991), pp. 165–86.

Meeks, W. A., 'The Image of the Androgyne: Some Uses of a Symbol in Earliest Christianity', *History of Religions*, 13:3 (1974), pp. 165–208.

Menéndez Pelayo, M., *Historia de los heterodoxos españoles*, 3 vols (Madrid: Librería Católica de San José, 1880).

Mercado, L., *De Mulierum Affectionibus* (Valladolid: Diego Fernández de Córdoba, 1579).

Mexía, P., *Silva de Varia Lección* (1540; Madrid: Cátedra, 1989).

Mitjavila, V., *Compendio de Policía Médica*, facs. edn by J. M. Calbet and J. Corbella (1803; Barcelona: Universidad de Barcelona, 1983).

Molina, F., 'Crónicas de la sodomía. Representaciones de la sexualidad indígena a través de la literatura colonial', *Bibliographica Americana. Revista Interdisciplinaria de Estudios Coloniales*, 6 (2010), pp. 1–12.

—, 'La *herejización* de la sodomía en la sociedad moderna. Consideraciones teológicas y *praxis* inquisitorial', *Hispania Sacra*, 62:126 (2010), pp. 539–62.

—, 'La sodomía a bordo. Sexualidad y poder en la Carrera de Indias', *Revista de Estudios Marítimos y Sociales*, 3:3 (2010), pp. 9–20.

Molina Artaloytia, F., 'Estigma e interacción: un análisis filosófico del discurso del Dr. Asdrúbal d' Aguiar sobre el homoerotismo', in A. L. Pereira and J. R. Pita (eds), *III Jornadas de História da Psiquiatria e Saúde Mental. Reunião Internacional* (Coimbra: Universidade de Coimbra-CEIS 20, 2012), pp. 7–12.

—, 'Los avatares (ibéricos) de la noción de sodomía entre la Ilustración y el Romanticismo', in F. Durán López (ed.), *Obscenidad, vergüenza, tabú: contornos y retornos de lo reprimido entre los siglos XVIII y XIX* (Cadiz: Universidad de Cádiz, 2012), pp. 101–20.

Monlau, P. F., *Higiene del Matrimonio o el Libro de los casados* (1853; Madrid: M. Rivadeneyra, 1868).

Montaña de Monserrate, B., *Libro de la Anathomía del Hombre* (Valladolid: S. Martínez, 1551).

Morant, I., and M. Bolufer Peruga, *Amor, Matrimonio y Familia. La construcción histórica de la familia moderna* (Madrid: Síntesis, 1998).

Morel d'Arleux, A., 'Las "Relaciones de Hermafroditas": dos ejemplos diferentes de una misma manipulación ideológica', in M. Cruz García de Enterría, H. Ettinghausen, V. Infantes and A. Redondo (eds), *Las Relaciones de sucesos en España (1500–1750). Actas del Primer Coloquio Internacional (Alcalá de Henares, 8, 9 y 10 de junio de 1995)* (Alcalá de Henares and Paris: Pub. Universidad Alcalá de Henares/Pub. Sorbonne, 1996), pp. 261–71.

Morgado García, A., 'El divorcio en el Cádiz del siglo XVIII', *Trocadero. Revista de Historia Moderna y Contemporánea*, 6–7 (1994–5), pp. 125–37.

—, *Ser clérigo en la España del Antiguo Régimen* (Cadiz: Universidad de Cádiz, 2000).

Mosácula, J., *Elementos de Fisiología Especial o Humana*, vol. 2 (Madrid: Hijos de C. Piñuela, 1830).

Moscoso, J., 'Monsters as Evidence: The Uses of the Abnormal Body during the Early Eighteenth Century', *Journal of the History of Biology*, 31:2 (1998), pp. 355–82.

Mott, L., 'Pagode português: a subcultura *gay* em Portugal nos tempos inquisitoriais', *Revista Ciência e Cultura*, 40:2 (1988), pp. 120–39.

—, and A. Assunção, 'Love's Labors Lost: Five Letters from a Seventeenth-Century Portuguese Sodomite', in K. Gerard and G. Hekma (eds), *The Pursuit of Sodomy: Male Homosexuality in Renaissance and Enlightenment Europe* (New York and London: Harrington Park Press, 1989), pp. 91–101.

Muñoz-Prián, S.; 'Identidades transgenéricas en la España del Antiguo Régimen. Un caso de cambio de sexo en la Andalucía del siglo XVIII' (unpublished MA thesis, Universidad de Cádiz, 2009).

Nederman, C. J., and J. True, 'The Third Sex: The Idea of the Hermaphrodite in Twelfth-Century Europe', *Journal of the History of Sexuality*, 6:4 (1996), pp. 497–517.

Nieremberg, J. E., *Curiosa y Oculta Filosofía* (1638; Madrid: Imprenta Real, 1643).

Nye, R., *Masculinity and Male Codes of Honor in Modern France* (New York and Oxford: Oxford University Press, 1993).

Oestreich, G., *Neostoicism and the Early Modern State* (Cambridge: Cambridge University Press, 1982).

Onaindía, M., *La construcción de la nación española. Republicanismo y nacionalismo en la Ilustración* (Barcelona: Ediciones B, 2002).

Opere, F., *Historias de la frontera: el cautiverio en la América hispánica* (Buenos Aires: FCE, 2001).

Orfila, M., *Tratado de Medicina Legal*, 4 vols (Madrid: Imprenta de Don José María Alonso, 1847).

Ortiz, T., 'From Hegemony to Subordination: Midwives in Early Modern Spain', in H. Marland (ed.), *The Art of Midwifery: Early Modern Midwives in Europe* (London: Routledge, 1994), pp. 95–114.

Ovid, *Metamorphoses*, trans. D. Raeburn (London: Penguin, 2004).

Padilla Manrique y Acuña, M. L. de, *Excelencias de la Castidad* (1642; Madrid: Biblioteca de Autores Españoles, 1975).

Pardo Tomás, J., *Ciencia y censura. La Inquisición Española y los libros científicos en los siglos XVI y XVII* (Madrid: CSIC, 1992).

Paré, A., *On Monsters and Marvels*, trans. J. L. Pallister (Chicago, IL and London: University of Chicago Press, 1982).

Park, K., *Doctors and Medicine in Early Renaissance Florence* (Princeton, NJ: Princeton University Press, 1985).

—, 'The Rediscovery of the Clitoris: French Medicine and the Tribade, 1570–1620', in D. Hillman and C. Mazzio (eds), *The Body in Parts: Fantasies on Corporeality in Early Modern Europe* (New York and London: Routledge, 1997), pp. 171–93.

—, 'Una historia de la admiración y del prodigio', in A. Lafuente and J. Moscoso (eds), *Monstruos y Seres Imaginarios en la Bilbioteca Nacional* (Madrid: Ministerio de Educación y Cultura, Biblioteca Nacional, 2000), pp. 77–90.

—, and L. Daston, 'Unnatural Conceptions: The Study of Monsters in Sixteenth and Seventeenth Century France and England', *Past and Present*, 92:1 (1981), pp. 20–54.

—, and R. A. Nye, 'Destiny is Anatomy', *New Republic* (1991), pp. 53–7.

Parker, G., *The Military Revolution: Military Innovation and the Rise of the West, 1500–1800* (Cambridge: Cambridge University Press, 1988).

Parsons, J., *Mechanical and Critical Inquiry into the Nature of Hermaphrodites* (London: J. Walthoe, 1741).

Passeron, J. C., *Le Raisonnement Sociologique. Un espace non -poppérien de l'argumentation* (Paris: Albin Michel, 2006).

Paterson, A. K. G., 'Tirso de Molina and the Androgyne: "El Aquiles" y "La Dama del Olivar"', *Bulletin of Hispanic Studies*, 70:1 (1993), pp. 105–13.

Peiró, P. M., and J. Rodrigo, *Elementos de Medicina y Cirugía Legal arreglados a la Legislación Española* (1832; Zaragoza: Imprenta de Mariano Peiró, 1841).

Peramato, P. de, *Opera Medicinalia: De elementis, De humoribus, De temporamentis* (Sanlúcar de Barrameda: Fernando Díaz Imp., 1576).

Pérez de Moya, J., *Philosophía Secreta* (1585; Madrid: Viuda de Alonso Martín, 1628).

Pérez de Petinto, M., and M. Bertomeu, 'Comienzo y actualidad (en 1951) de la trayectoria corporativa médico-forense', *Revista Española de Medicina Legal*, 23 (1999), pp. 5–43.

Pérez Escohotado, J., *Sexo e Inquisición en España* (Madrid: Temas de Hoy, 1992).

Pérez Ibáñez, M. J., *El Humanismo Médico del Siglo XVI en la Universidad de Salamanca* (Valladolid: Universidad de Valladolid, 1997).

Pérez Sánchez, A. E., J. Gallego and M. Mena, *Monstruos, enanos y bufones en la Corte de los Austrias* (Madrid: Fundación Amigos del Museo de Prado, 1986).

Perry, M. E., *Ni espada rota ni mujer que trota. Mujer y desorden social en la Sevilla del Siglo de Oro* (Barcelona: Crítica, 1993).

—, 'From Convent to Battlefield: Cross-Dressing and Gendering the Self in the New World of Imperial Spain', in J. Blackmore and G. S. Hutcheson (eds), *Queer Iberia: Sexualities, Cultures and Crossings from the Middle Ages to the Renaissance* (Durham, NC: Duke University Press, 1999), pp. 394–419.

Perym, D. de Froes, *Theatro Heroino, Vol. I: Abecedario historico, e catalogo das mulheres ilustres em armas, letras, acçoens heroicas, e artes liberaes* (Lisbon: Theotonio Antunes Lima, 1736).

Peset, J. L., *Ciencia y Marginación. Sobre negros, locos y criminales* (Barcelona: Crítica, 1983).

—, and M. Peset, 'Estudio preliminar', in *Lombroso y la Escuela Positivista Italiana* (Madrid: CSIC, 1975), pp. 80–1.

Pinto-Correia, C., and J. P. Sousa Dias, *Assim na Terra como no Céu. Ciência, Religião e Estruturação do Pensamento Ocidental* (Lisbon: Relógio D'Água, 2003).

Plenck, J. J., *Elementa Medicinae et Chirugiae Forensis* (1796; Madrid: Michaelis Burgos, 1825).

Pliny, *Natural History*, trans. H. Rackham, vol. 2, book 7 (London and Cambridge, MA: William Heinemann/Harvard University Press, 1942).

Pomata, G., 'Uomini Menstruanti. Somiglianza e Diferenza Fra i Sessi in Europa in Etá Moderna', *Quaderni Storici*, 27:1 (1992), pp. 51–103.

Proceso Inquisitorial de Elena o Eleno de Céspedes, Archivo Histórico Nacional, Inquisición, Legajo 234, n° 24.

Pulido, A. 'Lactancia Paterna', *Revista de Medicina y Cirugía Prácticas*, 7 (1880), pp. 13–22.

Rappaport, J., 'Mischievous Lovers, Hidden Moors and Cross-Dressers: Passing in Colonial Bogotá', *Journal of Spanish Cultural Studies*, 10:1 (2009), pp. 7–25.

Reis, E., *Bodies in Doubt: An American History of Intersex* (Baltimore, MD: Johns Hopkins University Press, 2009).

Restrepo-Gautier, P., 'Afeminados, hechizados y hombres vestidos de mujer: la inversión sexual en algunos entremeses de los Siglos de Oro', in M. J. Delgado and A. Saint-Saëns (eds), *Lesbianism and Homosexuality in Early Modern Spain* (New Orleans, LA: University Press of the South, 2000), pp. 199–215.

Révah, I. S., 'Les Marranes', *Revue des Études Juives*, 118 (1959–60), pp. 29–77.

Revel, J., 'El historiador y los papeles sexuales', in Various Authors, *Familia y Sexualidad en Nueva España* (México: FCE, 1982), pp. 53–4.

Rich Greer, M., 'María de Zayas and the Female Eunuch', *Journal of Spanish Cultural Studies*, 2:1 (2001), pp. 41–53.

Richards, J., *Sex, Dissidence, and Damnation: Minority Groups in the Middle Ages* (London and New York: Routledge, 1991).

Ricoeur, P., *Oneself as Another*, trans. K. Blamey (Chicago, IL and London, University of Chicago Press, 1992).

Riera Palmero, J., 'Andrés Laguna y el Galenismo Renacentista', in J. L. García Hourcade and J. M. Moreno Yuste (eds), *Andrés Laguna. Humanismo, Ciencia y Política en la Europa Renacentista* (Valladolid: Junta de Castilla y León, Consejería de Educación y Cultura, 2001), p. 166.

Rivilla Bonet, J., *Desvíos de la Naturaleza o Tratado del Origen de los Monstruos* (Lima: Imprenta Real, 1695).

Rodrigues, I. Teixeira, 'Amato Lusitano e as perturbações sexuais. Algumas contribuições para uma nova perspectiva de análise das *Centúrias de Curas Medicinais*' (unpublished doctoral thesis, Universidade de Trás-os-Montes e Alto Douro, 2005).

Rodríguez, J. C., *La literatura del pobre* (Granada: Comares, 1994).

Rodríguez, R. P. A. J., *Carta respuesta a un ilustre Prelado sobre el feto monstruoso hallado poco ha en el vientre de una cabra y reflexiones críticas que ilustran su historia* (Madrid: Imprenta Real de la Gaceta, 1753).

—, 'Disertación II. Sobre la imposibilidad de generación ni comercio por el Demonio íncubo', in *Nuevo Aspecto de Theologia Médico-Moral y ambos derechos o Paradoxas phísico-teológicas-legales*, vol. 2 (Madrid: Imprenta Real de la Gaceta, 1753), pp. 200–15.

Rodríguez, F. M. de, *Suma de casos de conciencia* (Salamanca: Imp. Diego de Cossío, 1603).

Rodríguez Ocaña, E., 'El resguardo de la salud. Administración sanitaria española en el siglo XVIII', in *Salud Pública en España. Ciencia, Profesión y Política, siglos XVIII–XIX* (Granada: Universidad de Granada, 2005), pp. 17–48.

Rodríguez Sánchez, A., *Hacerse nadie. Sometimiento, sexo y silencio en la España de finales del siglo XVI* (Lleida: Milenio, 1998).

Romera Navarro, M., 'Las disfrazadas de varón en la comedia', *Hispanic Review*, 2:4 (1934), pp. 269–86.

Ronzón, E., 'El Médico Juan Sánchez Valdés de la Plata y su libro sobre el Hombre. Historia de una investigación', *El Basilisco*, 24 (1998), pp. 63–84.

Rosen, G., *A History of Public Health* (1958; Baltimore, MD and London: Johns Hopkins University Press, 1993).

Rossell, A., *Manual de Medicina Legal* (Madrid: Ramón Rodríguez Rivera, 1848).

Rousseau, G. S., 'Nymphomania, Bienville and the Rise of Erotic Sensibility', in P. G. Boucé (ed.), *Sexuality in Eighteenth-Century Britain* (Manchester: Manchester University Press, 1982), pp. 95–119.

—, and R. Porter (eds), *Sexual Underworlds of the Enlightenment* (Manchester: Manchester University Press, 1987).

Roussel, P., *Sistema Físico y Moral de la Muger* (Madrid: Imp. de D. José del Collado, 1821).

Rubio Merino, P. (ed.), *La Monja Alférez Doña Catalina de Erauso. Dos Manuscritos Autobiográficos Inéditos* (Seville: Ediciones del Cabildo Metropolitano de la Catedral de Sevilla, 1995).

Rudelle-Berteaud, E., 'Divergencias moriscas y cristianas sobre erotismo y afectividad', *Al-Andalus*, 248 (2004), at http://old.webislam.com/numeros/2004/248/temas/divergencias_moriscas_cristianas.htm [accessed 4 May 2013].

S. N., 'Nueva aplicación del microscopio a los experimentos médico-legales', *Boletín de Medicina, Cirugía y Farmacia*, 2:66 (1841), p. 237.

Sabsay, L., 'De sujetos performativos, psicoanálisis y visiones constructivistas', in P. Soley-Beltran and L. Sabsay (eds), *Judith Butler en disputa. Lecturas sobre la performatividad* (Madrid and Barcelona: Egales, 2012), pp. 135–68.

Sabugosa, Conde de, *Neves de Antanho*, 3rd edn (Lisbon: Livraria Bertrand, 1919).

Saint-Hilaire, I. Geoffroy, *Histoire Générale et Particulière des anomalies de l'organisation chez l'homme et les animaux. Traité de Tératologie*, vol. 2 (Paris: J. B. Baillière, 1836).

Sáiz Carrero, A., 'Real Colegio de Cirugía de San Carlos', *Urología Integrada y de Investigación*, 14:2 (2009), pp. 188–206.

Salamanca Ballesteros, A., *Monstruos, Ostentos y Hermafroditas* (Granada: Universidad de Granada, 2007).

Sánchez, T., *De sancto matrimonii sacramento disputationum*, vol. 3 (1607; Avenione: Hyeronimum Duperier, 1689).

Sánchez Blanco Parody, F., *Europa y el Pensamiento Español del siglo XVIII* (Madrid: Alianza Universidad, 1991).

Sánchez Valdés de la Plata, J., *Crónica y Historia General del Hombre* (Madrid: Miguel Martínez, 1598).

Sánchez de Viana, P., *Anotaciones sobre los quinze libros de las Transformaciones, de Ovidio* (Valladolid: Diego Fernández de Córdoba, 1589).

Sánchez Vidal, A., *Esclava de Nadie* (Madrid: Espasa Calpe, 2010).

Sánchez Tortolés, A., *El Entretenido, primera parte, repartido en catorce noches* (Madrid: Lucas Antonio de Bedmar, 1673).

Santos, B. de Sousa, 'Between Prospero and Caliban: Colonialism, Post-Colonialism, and Inter-identity', *Luso-Brazilian Review*, 39:2 (2002), pp. 9–43.

Sanz Hermida, J., 'Aspectos fisiológicos de la dueña dolorida: la metamorfosis de la mujer en hombre', in *Actas del Tercer Coloquio Internacional de la Asociación de Cervantistas* (Barcelona: Anthropos, 1993), pp. 463–72.

—, 'Ensoñación y transformismo: la parodia erótica en *El sueño de la viuda*, de Fray Melchor de la Serna', in *Studia Aurea. Actas del Congreso de la AISO*, vol. 1 (Toulouse and Pamplona: Centro Virtual Cervantes, 1996), pp. 513–23.

—, 'La Literatura de problemas en España (siglos XVI y XVII)' (unpublished doctoral thesis, University of Salamanca, 1997).

Schiebinger, L., 'Skeletons in the Closet: The First Illustrations of the Female Skeleton in Eighteenth-Century Anatomy', in C. Gallagher and T. Laqueur (eds), *The Making of the Modern Body: Sexuality and Society in the Nineteenth Century* (London and Berkeley, CA: University of California Press, 1987), pp. 42–82.

—, *The Mind Has No Sex?: Women in the Origin of Modern Science* (Cambridge, MA and London: Harvard University Press, 1989).

—, *Nature's Body: Sexual Politics and the Making of Modern Science* (London: HarperCollins, 1993).

—, *Feminism and the Body* (Oxford: Oxford University Press, 2000).

—, 'Skelettestreit', *Isis*, 94:2 (2003), pp. 307–13.

Schleiner, W., 'Le feu caché: Homosocial Bonds between Women in a Renaissance Romance', *Renaissance Quarterly*, 45:2 (1992), pp. 293–311.

—, 'Cross-Dressing, Gender Errors, and Sexual Taboos in Renaissance Literature', in S. P. Ramet (ed.), *Gender Reversals and Gender Cultures: Anthropological and Historical Perspectives* (New York: Routledge, 1996), pp. 92–104.

Seed, P., *Ceremonies of Possession in Europe's Conquest of the New World, 1492–1640* (Cambridge: Cambridge University Press, 1995).

Shohat, E., *Dangerous Liaisons: Gender, Nations, and Postcolonial Perspectives* (Minneapolis, MN and London: University of Minnesota Press, 1997).

Simón Palmer, C., 'La higiene y la medicina de la mujer española a través de los libros (s. XVI al XIX)', in M. A. Durán et al. (eds), *La mujer en la historia de España (siglos XVI–XX)* (Madrid: Universidad Autónoma, 1984), pp. 71–84.

Smith, P. J., *The Body Hispanic: Gender and Sexuality in Spanish and Spanish American Literature* (Oxford: Oxford University Press, 1989).

Solomon, M., 'Fictions of Infection: Diseasing the Sexual Other in Francesc Eiximenis's *Lo Llibre de les dones*', in J. Blackmore and G. S. Hutcheson (eds), *Queer Iberia: Sexualities, Cultures and Crossings from the Middle Ages to the Renaissance* (Durham, NC: Duke University Press, 1999), pp. 277–90.

Soyer, F., 'The Inquisition and the "Priestess of Zafra": Hermaphroditism and Gender Transgression in Seventeenth-Century Spain', *Annali della Scuola Normale Superiore di Pisa. Classe di Lettere e Filosofia*, 5:1–2 (2009), pp. 535–62.

—, *Ambiguous Gender in Early Modern Spain and Portugal: Inquisitors, Doctors and the Transgression of Gender Norms* (Leiden and Boston, MA: Brill, 2012).

Steinberg, S., *La Confusion des Sexes. Le travestissement de la Renaissance à la Révolution* (Paris: Fayard, 2001).

Stepto, M., and G. Stepto (eds), *Lieutenant Nun: A Memoir of a Basque Transvestite in the New World: Catalina de Erauso* (Boston, MA: Beacon Press, 1996).

Stolberg, M., 'A Woman Down to her Bones. The Anatomy of Sexual Difference in the Sixteenth and Early Seventeenth Centuries', *Isis*, 94:2 (2003), pp. 274–99.

Sweet, J. H., 'Male Homosexuality and Spiritism in the African Diaspora: The Legacies of a Link', *Journal of the History of Sexuality*, 7:2 (1996), pp. 184–202.

Tellechea Idígoras, J. I., *La Monja Alférez, Doña Catalina de Erauso* (San Sebastián: Sociedad Guipuzcoana de Ediciones y Publicaciones, Obra Cultural de Kutxa, 1992).

Thoinot, L., *Tratado de Medicina Legal, traducido, anotado y adicionado con referencia a la legislación española y americana por W. Coroleu, Secretario Perpetuo de la Real Academia de Medicina de Barcelona*, vol. 2 (Barcelona: Salvat Editores, 1928).

Thorndike, L., *A History of Magic and Experimental Science, Vols 7 & 8: The Seventeenth Century* (New York and London: Colombia University Press, 1958).

Tomás y Valiente, F., 'El Crimen y Pecado contra Natura', in F. Tomás y Valiente et al., *Sexo Barroco y otras transgresiones modernas* (Madrid: Alianza Universidad, 1990), pp. 33–55.

—, 'Teoría y práctica de la tortura judicial en las obras de Lorenzo Matheu i Sanz (1618–1680)', in *La tortura en España* (Barcelona: Ariel, 1994), pp. 37–91.

Torre, E., 'Introducción', in J. Huarte de San Juan, *Examen de Ingenios para las Ciencias* (Madrid: Editora Nacional, 1976), pp. 9–45.

Tort, M., 'Le mixte et l'Occident. L'hermaphrodite entre le mythe et la science. Platon, Ovide, Isidore Geoffroy Saint Hilaire', in *La Raison Classificatoire* (Paris: Aubier Montaigne, 1989), pp. 175–203.

Tort, P., *L'Ordre et les Monstres* (Paris: Le Sycomore, 1980).

Valderrama, F., 'Si la muger que pare un Monstruo especie de Bruto, se deba presumir Reo de feo crimen por el Magistrado y como procederá contra ella', *Memorias Académicas de la Real Sociedad de Medicina y demás Ciencias de Sevilla*, 5 (1790), pp. 108–20.

Valencia García, M. A., *Simbólica femenina y producción de contextos culturales. El caso de la Santa Barbada* (Ávila: Institución Gran Duque de Alba, 2004).

Valverde de Amusco, J., *Historia de la Composición del Cuerpo Humano* (Rome: J. A. de Salamanca and A. Lafrery, 1556).

Vallbona, R. de (ed.), *Vida i sucesos de la Monja Alférez: Autobiografía atribuida a Doña Catalina de Erauso* (Tempe: Center for Latin American Studies, 1992).

Various Authors, *Diccionario de la Lengua Castellana en que se explica el verdadero sentido de las voces, su naturaleza y calidad*, vol. 3 (1732; Madrid: Ediciones Turner, 1977).

Vázquez García, F., 'Ninfomanía y construcción simbólica de la femineidad (España, siglos XVIII–XIX)', in C. Canterla (ed.), *VII Encuentro de la Ilustración al Romanticismo. La Mujer en los siglos XVIII y XIX* (Cadiz: Pub. Universidad de Cádiz, 1994), pp. 125–135.

—, 'La Exclusión del Hermafrodita y la Invención Ilustrada del Único Sexo Verdadero', in A. Romero Ferrer (ed.), *Actas del VI Encuentro de la Ilustración al Romanticismo. Juego, Fiesta y Transgresión (1750–1850)* (Cadiz: Servicio de Publicaciones de la Universidad de Cádiz, 1995), pp. 645–53.

—, 'La imposible fusión. Claves para una genealogía del cuerpo andrógino', in D. Romero de Solís, J. B. Díaz-Urmeneta Muñoz and J. López-Lloret (eds), *Variaciones sobre el cuerpo* (Seville: Servicio de Publicaciones de la Universidad de Sevilla, 1999), pp. 217–35.

—, 'Androginia y Pensamiento Esencial', *Culturas. Suplemento cultural del Diario de Sevilla*, 5 July 2001, p. 6.

—, *Tras la autoestima. Variaciones sobre el yo expresivo en la modernidad tardía* (San Sebastián: Gakoa, 2005).

—, 'Del Hermafrodita al Transexual. Elementos para una genealogía del cuerpo sexuado (España siglos XVI–XX)', in N. Corral (ed.), *Prosa Corporal. Variaciones sobre el cuerpo y sus destinos II* (Madrid: Talasa, 2008), pp. 75–97.

—, *La invención del racismo. Nacimiento de la biopolítica en España 1600–1940* (Madrid: Akal, 2009).

—, and R. Cleminson, 'Subjectivities in Transition: Gender and Sexual Identities in Cases of "Sex Change" and "Hermaphroditism" in Spain, *c.* 1500–1800', *History of Science*, 48:159 (2010), pp. 1–38.

—, and —, 'El Destierro de lo Maravilloso. Hermafroditas y Mutantes sexuales en la España de la Ilustración', *Asclepio. Revista de Historia de la Medicina y de la Ciencia*, 63:1 (2011), pp. 31–62.

—, and A. Moreno Mengíbar, 'Del Sexo Verdadero al Sexo Veraz. Reyes Carrasco, un caso de hermafrodismo en el siglo XIX', *El Viejo Topo*, 84 (1995), pp. 73–80.

—, and —, 'Un solo sexo. Invención de la monosexualidad y expulsión del hermafroditismo', *Daimón. Revista de Filosofía*, 11 (1995), pp. 95–112.

—, and —, 'El hermafrodita Reyes Carrasco. Identidad sexual en la España del siglo XIX', *Historia 16*, 258 (1997), pp. 30–6.

—, and —, *Sexo y Razón. Una genealogía de la moral sexual en España (siglos XVI–XX)* (Madrid: Akal, 1997).

—, and —, 'Hermafroditas y cambios de sexo en la España Moderna', in A. Lafuente and J. Moscoso (eds), *Monstruos y Seres Imaginarios en la Biblioteca Nacional* (Madrid: Ministerio de Educación y Cultura, Biblioteca Nacional, 2000), pp. 91–103.

Vega, C., 'Figuras de la exclusión femenina: ermitañas, monjas visionarias, barbudas, mujeres pilosas' (unpublished paper given at the I Curso del Aula de Salamanca, May 1990, Universidad de Salamanca).

Vega, V., *Diccionario Ilustrado de Rarezas, Inverosimilitudes y Curiosidades*, 4th edn (Barcelona: Gustavo Gili, 1971)

Velasco, S., *The Lieutenant Nun: Transgenderism, Lesbian Desire and Catalina de Erauso* (Austin, TX: University of Texas Press, 2000).

—, 'Marimachos, hombrunas, barbados: The Masculine Woman in Cervantes', *Cervantes: Bulletin of the Cervantes Society of America*, 20:1 (2001), pp. 69–78.

—, 'Interracial Lesbian Erotics in Early Modern Spain: Catalina de Erauso and Elena/o de Céspedes', in L. Torres and I. Perpetusa-Seva (eds), *Tortilleras: Hispanic and US Latina Lesbian Expression* (Philadelphia, PA: Temple University Press, 2003), pp. 213–27.

—, *Male Delivery: Reproduction, Effeminacy and Pregnant Men in Early Modern Spain* (Nashville, TN: Vanderbilt University Press, 2006).

—, *Lesbians in Early Modern Spain* (Nashville, TN: Vanderbilt University Press, 2011).

Vélez Quiñones, H., 'Deficient Masculinity: "Mi puta es el Maestre de Montesa"', *Journal of Spanish Cultural Studies*, 2:1 (2001), pp. 27–40.

Ventura Pastor, J., *Preceptos Generales sobre las Operaciones de los Partos* (Madrid: Joseph Herrera, 1789).

Vieira, A., *Arte de Furtar* (Lisbon: Livraria Peninsular Editora, 1937).

Vidal, D., *Cirugía Forense o arte de hacer las relaciones chirurgico-legales* (1783; Zaragoza: Imprenta de las Heras, 1814); facsimile edition by J. Corbella (Barcelona: Seminari Pere Mata, Universidad de Barcelona, 1987).

Vigarous, J. M. J., *Curso Elemental de las Enfermedades de las Mugeres*, 2 vols (Madrid: Imp. de Juan de Brugada, 1807).

Villacañas, J. L., 'El cosmos intelectual de Villalobos. Sobre el carácter de la primera modernidad hispana' (2012), at http://saavedrafajardo.um.es/WEB/archivos/NOTAS/RES0119.pdf [accessed 3 May 2013], pp. 1–38.

Villalobos, E. de, *Manual de Confesores* (Salamanca: Imp. Diego de Cossío, 1628).

Viñas y Mey, C., and R. Paz, *Relaciones Histórico-Geográfico-Estadísticas de los pueblos de España hechas por iniciativa de Felipe II. Provincia de Madrid* (Madrid: CSIC/Instituto Balmes de Sociología, 1949).

Virey, J. J., *Tratado Histórico y Fisiológico Completo sobre la Generación, El Hombre y la Muger* (Madrid: Imprenta de Antonio Martínez, 1821).

Wiesner-Hanks, M. E., *Christianity and Sexuality in the Early Modern World: Regulating Desire, Reforming Practice* (London: Routledge, 2000).

Wilson, D., *Signs and Portents: Monstruous Births from the Middle Ages to the Enlightenment* (London and New York: Routledge, 1993).

Yates, F. A., *The Rosicrucian Enlightenment* (London and New York: Routledge, 2002).

Yerushalmi, Y. H., *From Spanish Court to Italian Ghetto. Isaac Cardoso: A Study in Seventeenth-Century Marranism and Jewish Apologetics* (New York and London: Columbia University Press, 1971).

Zambrano, M., *Persona y democracia. La historia sacrificial* (1958; Madrid: Ediciones Siruela, 1996).

Zamora Calvo, M. J., '*In virum mutata est.* Transexualidad en la Europa de los siglos XVI y XVII', *Bulletin Hispanique*, 110:2 (2008), pp. 431–47.

Zarzoso Orellana, A., 'Policía y Ciencia de la Policía en el Discurso Urbanístico a Finales del Antiguo Régimen', *Asclepio*, 50:1 (1998), pp. 259–64.

—, 'La Pràctica Mèdica a la Catalunya del segle XVIII' (doctoral thesis, Universidad de Barcelona, 2003).

Zemon Davis, N., *Il Ritorno di Martin Guerre. Un caso di doppia identità nella Francia del Cinquecento* (Turin: Einaudi, 1984).

INDEX